# 100 THINGS TO SEE IN THE NIGHT SKY

**EXPANDED EDITION**

# 100 THINGS TO SEE IN THE NIGHT SKY

## EXPANDED EDITION

**YOUR ILLUSTRATED GUIDE TO THE PLANETS, SATELLITES, CONSTELLATIONS, AND MORE**

DEAN REGAS

ADAMS MEDIA

NEW YORK  LONDON  TORONTO  SYDNEY  NEW DELHI

Adams Media
An Imprint of Simon & Schuster, Inc.
57 Littlefield Street
Avon, Massachusetts 02322

For information about special discounts for bulk purchases, please contact Simon & Schuster Special Sales at 1-866-506-1949 or business@simonandschuster.com.

The Simon & Schuster Speakers Bureau can bring authors to your live event. For more information or to book an event contact the Simon & Schuster Speakers Bureau at 1-866-248-3049 or visit our website at www.simonspeakers.com.

Interior design by Colleen Cunningham and Julia Jacintho
Interior illustrations by Eric Andrews
Photographs © 123RF/Andrey Armyagov, Viktar Malyshchyts, fabio formaggio, Andrey Nyrkov; Getty Images/aapsky, allou, Cylonphoto, davidhajnal, Eerik, Elen11, ewg3D, FlashMyPixel, GRAWLLF, hadzi3, heibaihui, ilbusca, incposterco, juan lópez, leolintang, Martin Holverda, Mike Ries, Morrison1977, Muhammad Abu Dzar Al Ghifari, Nick Pandevonium, parameter, Paul Wilson, Riekkinen, seebest, shaunl, Sjo, standret, Stocktrek Images, wisanuboonrawd, Xiong Yi; NASA/ NASA/Aubrey Gemignani, USGS, JPL-Caltech/ Space Science Institute, NASA's Scientific Visualization Studio

Manufactured in the United States of America

4 2021

Library of Congress Cataloging-in-Publication Data has been applied for.

ISBN 978-1-5072-1381-0
ISBN 978-1-5072-1382-7 (ebook)

Contains material previously published in the following title published by Adams Media, an Imprint of Simon & Schuster, Inc.: *100 Things to See in the Night Sky* by Dean Regas, copyright © 2017 by Dean Regas, ISBN 978-1-5072-0505-1.

# Contents

# THE 100 THINGS TO SEE IN THE NIGHT SKY

# Introduction

*Constellations. Stars. Planets. Nebulas. Satellites.*

The universe is full of wonders beyond our imagination. And while much of what's out there in space can only be seen by powerful telescopes or theorized about by professional astronomers, there are many things you can see in the night sky just by walking out your door and looking up.

In *100 Things to See in the Night Sky* you'll find 100 entries, each telling you about particular objects in the night sky that you can find with either your naked eye, binoculars, or a backyard telescope. Each entry tells you what type of heavenly object you are looking for (planet, star, etc.), how hard it is to find, what the item looks like and its mythology, and where and when to look for it. You'll also find a variety of star maps scattered throughout the entries that will show you exactly where to look when you turn your eyes toward the stars.

The tips, techniques, and informative bits that you'll find in the book are the same ones that I used when I first started heading outside every clear night to let the sky become my classroom. They're the same tips that turned me into a starstruck astronomer. Hopefully they'll do the same for you, and help you fall in love with the heavens!

But what exactly will you find in this book? Well, first you'll learn how to safely observe the Sun, take in sunsets and sunrises, chart the seasons, and observe sunspots. Then you'll explore the phases and features of the Moon and uncover the hidden secrets of the five planets closest to Earth: elusive Mercury, dazzling Venus, ruby-red Mars, giant Jupiter, and spectacular Saturn.

Then you will delve into the heart of the book and study the art of stargazing. You will learn how to identify major stars and constellations throughout every season of the year. From beginner star patterns like the Big Dipper and Orion's Belt to more challenging constellations like Delphinus, the Dolphin, and Aries, the Ram, you will soon be able to recognize dozens of stars in the night sky and retell their ancient mythological stories.

Finally, you'll discover the tricks to observing man-made satellites and supreme heavenly shows such as meteor showers and eclipses.

And, in case you're not sure exactly what you need to do, you'll find a section packed with information on ideal stargazing locations and sky conditions, what type of equipment you need (in some instances, you just need to use your two eyes and bare hands), the best time to stargaze, and more. So, whether you're looking to learn about the night sky on your own or are looking to get your kids excited about what lies in the skies above, this book is for you. I challenge you to go outside and find every one of the night sky objects detailed here. When you do that, you may be converted to the stars like me, and you may find that you're an astronomer too!

# How to Use This Book

The universe is yours to behold and this book can be your beginner's guide. It will focus on 100 of the most amazing astronomical objects that you can see in the night sky. You'll find basic information and tips to locate each individual object, often with the help of accompanying charts and graphics. But some general rules and guidelines are necessary to start properly. Let's talk about viewing locations, sections of the sky, various viewing conditions, and any equipment you may have or want to get.

## LOCATION, LOCATION, LOCATION

This book is intended mainly for viewing audiences in:

- The mainland United States
- Southern Canada
- Northern Mexico
- Europe
- North Africa
- Most of Asia

It is designed to help you view the night sky from the mid-northern latitudes, which means that if you're living or traveling between 25 and 55 degrees north latitude, this is the book for you. Your perspective on the heavens does not change much when you travel east to west, but it does change when you trek north or south. There will be some useful information in this book no matter where you live, but my observing tips and star charts are mainly geared for stargazers living in the mid-northern latitudes.

### Break Up the Sky

That said, no matter where you are on Earth, if you stargaze long enough you will notice that the stars, constellations, planets, and the Sun and Moon slowly move across the sky. Hour by hour, minute by minute, they shift as one body. Ancient astronomers pictured the dome of heaven circling around an unmoving Earth, as if the gods were manipulating a great wheel behind the scenes to make everything rise and set and circumnavigate the globe once a day.

For beginning stargazers to really experience this motion it helps to break up the sky by the four cardinal directions: north, south, east, and west. When you face north and watch the stars, you will see that they behave quite differently than when you face any other direction. Take some time to sit outside under the stars and note the positions of several bright stars all around you and place them in reference to landmarks like houses, trees, and mountains.

Knowing how things move in each direction will help you know what stars and constellations are coming next and which objects might be setting soon.

If you live in mid-northern latitudes (between 25 and 55 degrees north latitude), on every night of the year this is what you will notice:

### NORTH

When you face north, the stars will seem to move in a counterclockwise circle around Polaris, a.k.a. the North Star. You can see star patterns like the Big Dipper and the W-shaped Cassiopeia arcing around the North Star. Many of the stars nearest to the North Star and the constellations in the northern sky never rise or set but endlessly seem to circle. Astronomers call these circumpolar stars.

### WEST

When you face west, the motion seems quite different. Stars, planets, the Moon, and the Sun appear to move diagonally down and to the right until they set below the horizon.

### EAST

When you face east everything rises and travels diagonally up and to the right.

### SOUTH

When you turn south, the celestial objects move from left to right and almost crawl across the southern horizon.

If you are in the Southern Hemisphere, your motions seem different. Objects still rise in the east and set in the west, but they travel up and to the left and down and to the left respectively. Objects in the northern sky move right to left, and the stars in the southern sky completely circle the pole in a clockwise direction. No matter where you live, all of these apparent motions are caused by one thing: the daily rotation of the Earth.

## PRIME-TIME STARGAZING

You probably won't be surprised to hear that more people watch the stars during the prime-time hours (around 8 or 9 p.m.) than any other time of day. The observing suggestions throughout the book for both the stars and constellations usually reflect this timing. So, for example, when I describe a group of stars as winter constellations or summer constellations, I mean that they are observable during the evenings of their respective seasons. That said, please note that seasonal stars can be seen in other months, but at different times of night. For example, you can see Orion in the winter evenings, but you can also find him gracing the sky on summer mornings just before dawn.

## SKY CONDITIONS

Obviously if you're going to head outside to stargaze, you'll want to pay attention to what's going on in the sky. The first thing you need to know is that clouds are the bane of the astronomer's existence. Avoid them at all costs. Whether you're viewing with your naked eye, telescope, or binoculars, you will need clear skies to see the maximum number of stars.

You also need to take light pollution into consideration. As cities grow larger so do their domes of light. The more light that we shine upward, the less light that the stars shine down. From urban locations you can sometimes barely make out the brightest twenty to thirty stars in the sky, called the first magnitude stars. But the farther you travel away from light pollution the more stars you can see. In the suburbs you may see second, third, and fourth magnitude stars, which include the brightest 700 stars in the sky. In a truly dark sky you can see the fifth and even sixth magnitude stars to behold up to 6,000 stars. So in the countryside, you can sometimes observe stars with the naked eye that are 100 times fainter than the stars people see in large cities. Additionally, from a dark sky you can see the Milky Way and even a few other deep sky objects that are between 1,000 and 2.5 million light-years away.

Most of the objects described in this book are first or second magnitude objects and are easily visible from light-polluted or semi-light-polluted skies. Throughout the book, you'll find info on how easy (or not-so-easy) it is to see these night sky objects. The first and second magnitude objects are labeled as Easy; third and fourth magnitude objects are labeled as Moderate; and anything that's harder to see than that is labeled Difficult. Such Difficult objects include the Hercules Cluster and the Andromeda Galaxy. Hopefully these designations will spur you to seek out the darkest locations so that you can see a sky full of stars.

The Moon itself can actually negatively affect your stargazing. While a big, beautiful Full Moon may add to the romance of an evening, it also puts out a lot of light. This added moonlight can wash out the fainter light of the stars and limit the number of objects you can detect. For the optimal stargazing experience, plan your observations around the New Moon and crescent phases.

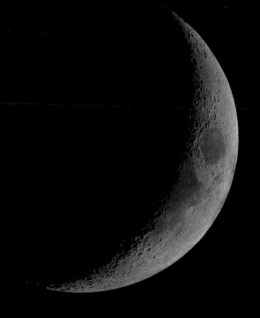

## EQUIPMENT

When you are trying to find faint objects in the sky with the naked eye, it may take you a while to see them. Even if you are away from light pollution and clouds, there is no Moon in the sky, and you have ideal viewing conditions, you still might have trouble finding some of these naked-eye objects. Do not be deterred. The objects are still there and can usually be seen without any additional equipment. However, if you're struggling to find them then using a pair of binoculars or peering through a telescope can help you enhance the experience.

You see, the more light you can gather, the more you can magnify an image and the more details you can see. And that is precisely what binoculars and telescopes do. With the exception of viewing the Sun, if you have binoculars or a telescope try pointing them at every object in this book for easier viewing. (Note: In this book I will tell you how to safely view the Sun.)

### Use Your Hands to Find Your Angles

Amateur and professional astronomers break up the sky into degrees and use them to determine how high an object is in the sky or the angular separation between two objects (how far apart they seem in the sky). In order to help you locate objects, you'll find references to degrees in the entries throughout the book. Conveniently, you have the built-in tools you need to measure angles yourself.

| 180° | 90° | 45° | 30° |
|---|---|---|---|
| Picture the sky as a dome or hemisphere. This gives you 180 degrees of sky to work with—from horizon to horizon. | If an object is straight overhead (the zenith), it is 90 degrees above the horizon. | A star halfway up the sky is therefore at 45 degrees above the horizon. | A star that is one third of the way above the ground is at about 30 degrees above the horizon. |

To make finer angular measurements, use your hand at arm's length.

- If you extend your arm and make a fist, that fist will cover approximately 10 degrees of the sky.
- If you open your hand and spread out your fingers, the space between the end of your thumb and the tip of your pinkie will measure about 25 degrees.
- The width of your pinkie at arm's length is only 1–2 degrees.

These measurements are not perfect, but making rough estimates like these can help you hop from a constellation you know to one you want to find. This method can also help you locate fainter stars that may be near brighter ones.

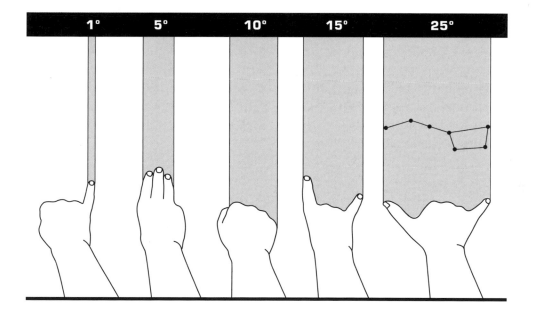

## PATIENCE AND PRACTICE

The best teacher of astronomy is the sky itself. Get outside every clear night and watch and learn. If you do this regularly, after less than 1 month the sky above will all start to make sense. You'll not only come to know the major stars and constellations, the planets, Moon, and Sun but will develop an appreciation for the heavens above. Let's get started!

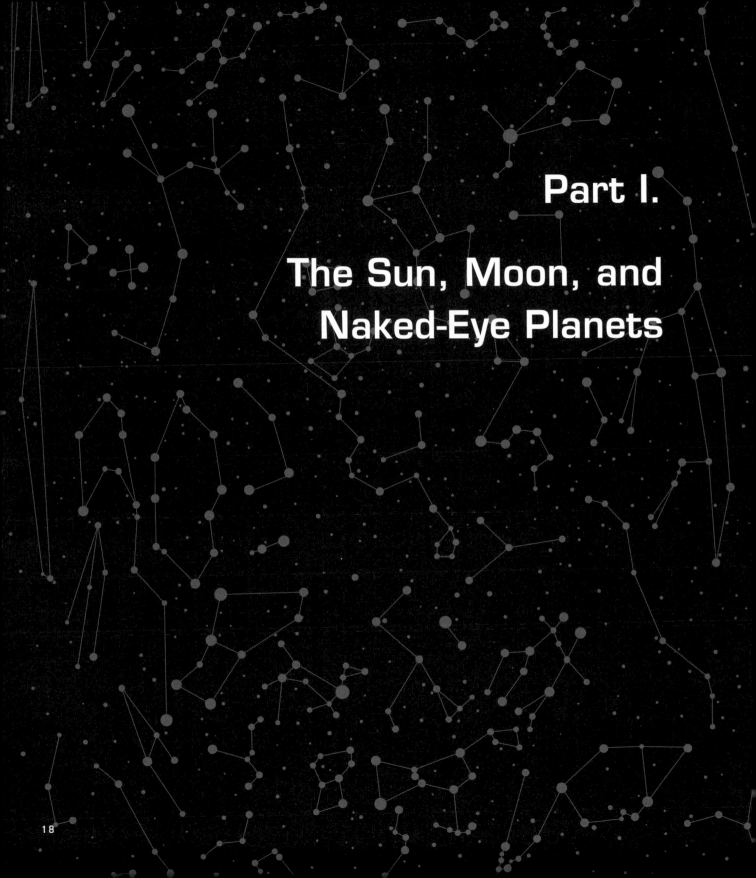

# Part I.

# The Sun, Moon, and Naked-Eye Planets

How do you get started studying the universe? I recommend doing what the ancients did: Observe the brightest things first. In this part of the book, we will explore the brightest, most noticeable objects in the day and night sky.

This part starts off with the Sun. I know it may seem funny to begin talking about the brightest thing in the daytime sky, but there are so many cool aspects of the Sun that you should check out more closely. Observing the daily motion of the Sun taught our ancestors about time. Monitoring its changing path from month to month taught them the seasons. And now modern astronomers can peer deeply through the sunlight and tell us how hot the Sun is, how far away it is, and what it is made of.

Here you'll also learn about the Moon, the second brightest object in the sky. Then we will fly to the five planets that are visible to the naked eye: Mercury, Venus, Mars, Jupiter, and Saturn. Venus shines in as the third brightest celestial object, with Jupiter and Mars coming in at numbers four and five, respectively. Although there are several stars that can be brighter than Mercury and Saturn, these planets make such interesting sights—much different than stars around them.

So let's get started and take a look at the objects detailed in this part. Each of them—the Sun, Moon, and the five naked-eye planets—are all so bright and so stunning that they will be difficult to miss.

# 01 | THE SUN

CLASSIFICATION: **OUR CLOSEST STAR** | VISIBILITY: **EASY**

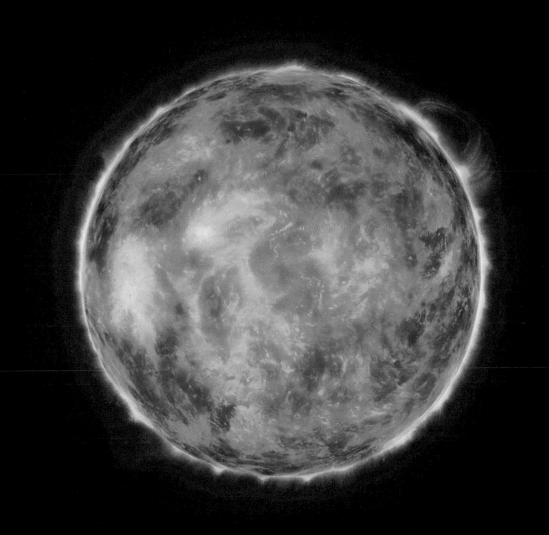

The Sun is our nearest star (about 93 million miles from Earth) and is the major source of light and heat for all the planets, moons, and asteroids in the solar system. At roughly 865,000 miles wide, the Sun is about 1,300,000 times larger than Earth. It is so massive that it holds all of the planets in steady orbits.

## SUN SAFETY

Whenever you view the Sun, think before you look. Do not look at the Sun for more than a moment unless you are 100 percent sure you are using safe and proven viewing methods. Never use common homemade filters including CDs, smoked glass, exposed film, X-ray transparencies, Pop-Tarts wrappers, or Mylar balloons. Your vision can be permanently damaged with even short periods of direct exposure to sunlight.

The best way to directly observe the Sun safely with the naked eye is by looking through either a sheet of #14 welder's glass or specially made eclipse glasses. You can get #14 welder's glass at welding supply stores, and several companies sell effective eclipse glasses online. Costing between $1 and $5, these items will allow you to observe the Sun's disc, partial eclipses, and maybe even a large sunspot. If you use the welder's glass, be sure to use only #14. That is the only shade dark enough to protect your eyesight—anything less is unsafe.

You'll notice that the Sun seems surprisingly small in the sky when you examine it by looking through #14 welder's glass or eclipse glasses. The Sun is humongous, but after you block out most of the surrounding glare, it looks like a tiny circle of light. After all, you are looking at the Sun from 93 million miles away! It's not exactly close to you.

You can also take pictures of the Sun through a sheet of welder's glass or through eclipse glasses:

- Place the welder's glass or eclipse glasses in front of a camera.
- Aim the camera at the Sun and review the image on your screen.
- Center the image, adjust the contrast, and snap away.

You'll have pictures of the Sun!

The welder's glass or eclipse glasses act like a filter that will reduce the glare and will protect your camera in the same way it protects your eye.

## SEE FEATURES ON THE SUN

Now that you have your safe solar-viewing glasses, what can you see? To the naked eye, on a normal day, you can't make out a lot of detail on the Sun. But sometimes you can pick out larger sunspots or, on very rare occasions, a solar eclipse.

Sunspots are darker, cooler regions on the Sun where magnetic disturbances cause stellar material to erupt from the surface. Through safe glasses, they look like tiny black blemishes on the glowing disc of the Sun. Only large sunspots can be detected with the naked eye, and finding them can be a test of your eyesight. They are more frequent during solar maximum, which is a varied period of time that occurs about every 11 years, when solar activity ramps up. But sunspots can pop up at any time, so it is always worth taking a look. If you are lucky enough to see a sunspot with the naked eye, remember that the spot is larger than Earth!

The best time to observe the Sun is during a solar eclipse. The most common eclipse is a partial solar eclipse, which is when the Moon covers only a portion of the solar disc. It looks like someone took a bite out of the Sun. Although solar eclipses occur about twice a year, they are localized events, which means each eclipse is visible only from certain areas on the globe. For any one location, you may see a solar eclipse about once every 3 to 6 years.

The best upcoming solar eclipses that will be visible from most parts of the United States will occur on:

- October 14, 2023
- April 8, 2024
- January 14, 2029

Use your eclipse glasses to get a safe view of these cool alignments of Sun, Moon, and Earth and admire the fact that astronomers, after watching the heavens for centuries, can now predict them well in advance!

### SEE SEVERAL SUNRISES AND SUNSETS

Even though you experience one sunset and one sunrise every day, you may not have considered the complex series of steps that happen during these events. To start, you want to look for several subtle changes around the sky in addition to the Sun.

#### Sunrises

For a sunrise, simply get up while it is still dark and then watch as the sky slowly brightens. While the Sun is almost ready to rise in the east, look behind you to the west. You will see a slightly darker piece of sky just above the horizon. That is Earth's shadow. When you look back to the east again you will notice every layer of the atmosphere igniting in different warm tones. The rising but still unseen Sun turns the atmosphere closest to the horizon from deep blue to ruby red in a matter of minutes. The wind may change as the temperature rises in anticipation of the rising Sun. And then, if you have a clear view to the horizon, you will see the top of the Sun peek up above it. It will take a few minutes for the Sun to fully clear itself from the horizon, but when it does then day has broken.

#### Sunsets

Experiencing a sunset is just as powerful (and you don't have to get up extra early). The play of light and darkness is reversed as you face west to see the Sun slowly set. The sky gradually turns from light to dark through every color in the rainbow. Just before sunset, turn around and face east to see the shadow of the Earth cast on the lower reaches of the atmosphere. Then face west again and bask in the final rays of sunlight as the Sun dips below the distant horizon. Night has fallen.

### SENSE THE SEASONS

You may have heard that "the Sun rises in the east and sets in the west." Well, that's mostly accurate. If you watch several sunrises or sunsets over the course of a few months, you will notice that the Sun does not always rise or set in the same place. And you may have seen that the Sun is much higher in the sky during the summer months than it is in the winter. Observing these changes in the position of the Sun is one of the oldest astronomical practices and can be done very easily with the naked eye.

In the following diagram you can see how the Sun moves across the sky in different seasons from any location in the mid-northern latitudes.

- On the winter solstice (the bottommost arc on the diagram) the Sun rises south of east, cuts low across the southern sky, and then sets south of west.
- Three months later, on the spring equinox (the middle arc of the diagram), the Sun rises due east, reaches a higher altitude in the south, and sets due west. Notice that the lengths of the Sun's arcs are different—in spring the Sun rides higher in the sky and stays visible for a much longer period of time than it does in the winter.
- On the summer solstice (the topmost arc of the diagram) the Sun rises north of east, travels extremely high in the southern sky, and then sets north of west. This is called the longest day because you experience the most minutes of daylight and the least amount of darkness for the year.

Sketch what you see along the horizon; note and date the changes throughout the year. You will quickly learn that we need to modify the rule to instead read, "The Sun rises east-ish and sets west-ish."

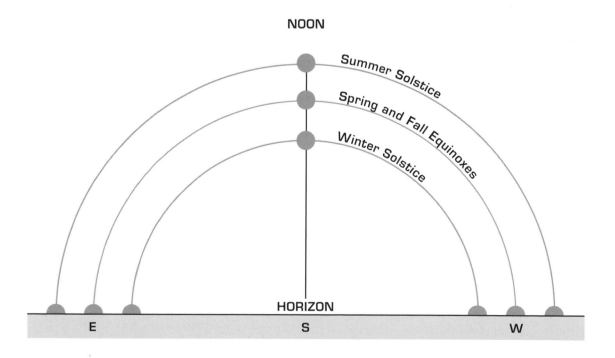

# 02

# THE MOON

| CLASSIFICATION: EARTH'S NATURAL SATELLITE | VISIBILITY: EASY |

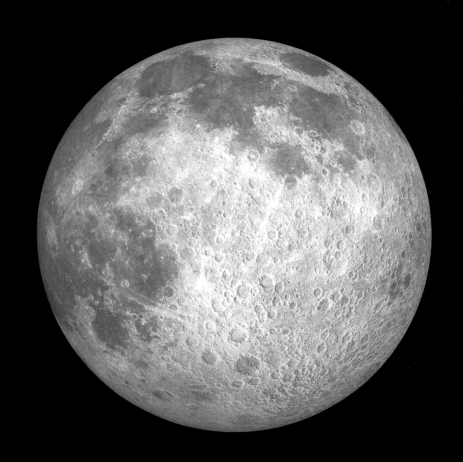

For thousands of years our ancestors have gazed up at the silvery glow of the Moon and tried to make sense of its rhythm and aura. What are the dark and light portions of the Moon? What's a Harvest Moon? And why does the Moon change shape and size? Memorialized in literature, song, and dance since the dawn of time, the Moon tugs at our romantic nature and inspires our soul.

The Moon is our closest neighbor in space: a ball of rock 2,159 miles in diameter that circles Earth from a distance of about 240,000 miles. It is the easiest and most dynamic object in the sky to view with the naked eye. Not only can you observe the Moon in the night sky during parts of each month, you can often catch sight of it during the daytime. It changes positions from night to night and rises and sets in different places regularly. And for a few days each month, you can't see it at all.

Waning Crescent

Waning Gibbous

Third Quarter

New

Full

First Quarter

Waxing Crescent

Waxing Gibbous

## OBSERVE THE PHASES OF THE MOON

You've noticed that the Moon doesn't look the same every time you look up into the night sky. Each night, it goes through a different phase. The chart on this page illustrates these phases. So what causes all these shapes? The answer is light! The Moon generates no light of its own—it shines with reflected sunlight. Light travels 93 million miles from the Sun, bounces off the surface of the Moon, and then travels another 240,000 miles to reach your eyes. The phase you see depends on where the Moon is in its orbit around Earth. When the Moon is between the Sun and Earth the only part of the Moon that is lit faces the Sun. This is the New Moon phase, and from Earth the Moon appears completely dark. When the Sun, Moon, and Earth form a right angle, you will see only half of the Moon lit (that happens at first and last quarter). And when the Moon is on the opposite side of Earth with respect to the Sun, you will see the entire Moon illuminated. That's a Full Moon.

It takes the Moon about 29.5 days to go through its entire cycle of phases, and so every night the Moon's phase changes slightly as different parts of the lunar surface are lit by the Sun, while others plunge into darkness.

Some people incorrectly believe that the phases of the Moon are caused by Earth casting a shadow on it. That makes sense when you see a crescent Moon, but how can a round Earth create a straight-line shadow like with a first quarter Moon, or the curve on a gibbous Moon? Earth can't make those phases of the Moon that you can see all month.

Observe the Moon every clear night during a month and check out not only how the phase changes but how its location in the sky and time that it is visible changes. Remember, to complete its orbit, each night the Moon moves about one-twenty-ninth the way around Earth. So it will appear to shift farther to the east 12–14 degrees from night to night. Although that's just a little larger than the width of your fist at arm's length, the shift is fast (astronomically speaking). This motion of the Moon around Earth also causes the Moon to rise about 50 minutes later each night. That time varies depending on the season and the phase of the Moon, but test it out. Make a Moon journal and sketch the position of the Moon every day at the same time. Soon you may be able to predict where and when the Moon will travel in the future.

### DRINK IN SOME EARTHSHINE

Have you ever looked at a crescent Moon and noticed that the dark part of the Moon looks like a faintly glowing, gray ball? This effect is caused by earthshine, where the light of the Sun shines on the Earth and reflects up on the Moon. Earthshine brightens the Moon enough so that you can see all of it against the darkness of the sky. Look for this effect during either a waxing crescent Moon (visible just after sunset) or a waning crescent Moon (visible just before sunrise).

### EXPERIENCE THE MOON ILLUSION

Have you noticed that the Moon looks bigger when it is near the horizon? Well, although the Moon appears larger near the horizon it really isn't. This effect is called the Moon illusion and scientists have debated about it for years. Astronomers first thought that the Moon appeared larger on the horizon because we have reference points like trees and buildings. But now they believe that the Moon illusion is caused by how we perceive the sky. We picture space on the horizon to be farther from us than overhead. Since the Moon

is actually the same size, the "farther" Moon becomes larger in our minds. The illusion takes place in our brains. You can combat the Moon illusion by looking at it upside down, between your legs. For some reason the Moon will look normal-sized. Try it. It might be embarrassing to do, but it really works.

Whenever you see the Moon low on the horizon, you can measure its apparent size in the sky. Just hold a hole-punched index card at arm's length. The Moon should be about the same size as the hole. Later at night measure the Moon again when it's higher in the sky. It will still be the same size as the hole!

### SEE A SUPERMOON!

The Moon's orbit around Earth is not a perfect circle. It travels in a slightly stretched oval shape, called an ellipse. This elliptical path means the Moon changes its distance from Earth. When the Moon is slightly closer to Earth, it can appear slightly larger in the sky. The closest Full Moon in a calendar year is known as a Supermoon. A naked-eye observer can't differentiate the Moon's apparent size from night to night. However, when one compares a Supermoon to the farthest Full Moon, a so-called Puny Moon, the variance is dramatic. The Supermoon is more than 31,000 miles closer to Earth, and consequently it appears 14 percent larger in diameter with a 30 percent larger surface area than the Puny Moon. That is like comparing a 16-inch pizza to a 14-inch pizza, or the size of a quarter to a nickel.

In the second century B.C. the Greek astronomer Hipparchus noticed this changing Moon size with his naked eye. He constructed a device called a diopter, a two-meter long stick with a sighting circle on the far end that could help him measure very small angles in the sky. After utilizing the diopter through several lunar cycles, Hipparchus discovered that the Moon's angular size (how large it appears to be in the sky) changed. Any time you observe a Full Moon, see if you can tell the difference between a Supermoon and a Puny Moon.

## SAILING THE SEAS AND SCALING THE HIGHLANDS ON THE MOON

There are two main terrains on the Moon that you can see with the naked eye: The brighter areas are called the highlands, and the darker areas are called the seas, or maria. The highlands are older structures and include many mountain peaks and a large number of craters. But your eye will also be drawn to the darker seas that cover 30 percent of the lunar surface.

When you observe a Full Moon, notice that the seas are mostly circular in shape. Each sea is really the remnant of a gigantic crater. Eons ago, the Moon was hit so hard by comets and meteors that it literally cracked. Over time, magma seeped out of these cracks and filled in the craters to form the seas. Some of these seas bear features that have given rise to their peaceful names, like the Sea of Serenity and Sea of Nectar. Others have more ominous titles, such as the Sea of Crises and Sea of Rains. The most famous lunar sea is the Sea of Tranquility. Neil Armstrong and Buzz Aldrin planted the first human footprints there on July 20, 1969.

The outlines of the highlands and the seas can often spark your imagination, and if you look long enough you may start to see shapes and patterns in these splotches on the face of the Moon. And maybe you will imagine an actual face on the surface of the Moon. At certain Moon phases, the play of light and shadow on the lunar surface may inspire you to see the "Man in the Moon." With a little more imagination you can connect the seas in such a way as to form a rabbit with two long ears and a fluffy tail. Or maybe it will look like a man carrying a bucket of water or perhaps a lady wearing a necklace. When you gaze at the Moon, let your imagination run wild.

## OBSERVING THE MOON WITH BINOCULARS AND TELESCOPES

With a little magnification you can start to see more details and features on the Moon. Even a simple pair of binoculars can provide an amazing view of the lunar surface. Most binoculars can be attached to a tripod with the aid of a simple adapter. If you can mount your binoculars on a steady tripod, you can watch the Moon for hours.

Round craters, high mountains, and deep valleys are all on display whenever you look at the Moon through a small telescope. Observing the Moon through a telescope is best done during the crescent, quarter, or gibbous phases (during a Full Moon, the surface is too bright and displays less fine detail). You'll want to focus your attention on the dividing line between the light and dark portions of the Moon. This line is called the terminator and that is where you will want to look to see the most dramatic plays of light and shadow on the uneven lunar landscape. The Moon's surface may look totally fake—like it is a bad movie set or made of clay. But that is the real Moon, up close and personal.

# The Naked-Eye Planets

The ancients had a much different idea about the planets than we do today. They could not picture them as round rocky or gaseous objects that circle the Sun. They were not worlds or places that could be visited. They were simply unique pinpoints of light that shone from an unfathomable distance. Or else they were gods. Without a telescope, there was no way for an ancient astronomer to get a closer look at the planets, but there was one characteristic that made the planets totally fascinating: They wandered.

To the naked eye of these ancient stargazers, the planets seemed to be very strange moving, sometimes suspiciously bright stars. In fact, the word *planet* comes from the ancient Greek word *planetes*, which means "wandering star." While all of the stars remained fixed in their recognizable formations, there were seven exceptions to this rule. They wandered from place to place over days, months, and years.

We have already talked about two ancient "planets": the Sun and Moon. Since the Sun and Moon both appeared to wander across the stars, they were considered to be planets by most ancient civilizations. The Sun moved slowly (about 1 degree per day), while the Moon took big jumps from night to night (about 13 degrees per night). But there are five other wandering planets that amazed, perplexed, and inspired the ancient stargazer—the five planets that you can see with the naked eye: Mercury, Venus, Mars, Jupiter, and Saturn.

Let's observe these five naked-eye planets through the eyes of the ancients. What made the planets unique? How did they move? What characteristics did each planet evoke? Then let's take some closer-up views of these wondrous objects and share something every ancient astronomer would be envious to see: a view of each planet through a small telescope.

Each planet has a personality. Each wandering star wanders through the constellations in a unique fashion. Mercury, associated with the swift, fleet-footed messenger god, lives up to the reputation of its namesake. It is the fastest planet, whipping around the Sun at an average speed of almost 106,000 miles per hour.

Mercury is the most elusive planet and is probably the toughest of the five naked-eye planets to find in the night sky. It moves quickly from night to night, and it shifts from being visible in the morning sky to the evening sky in a matter of weeks. It took a long time for the ancients to figure out what was going on up there. In fact, some cultures considered Mercury to be two separate objects—a morning deity and an evening deity.

## BEST TIMES TO SEE MERCURY
The main problem with finding Mercury is that it orbits very close to the Sun. That means Mercury is most often up in the sky at the same time as the Sun. Sunlight washes out Mercury's feeble light, and this makes it invisible to the naked eye during most of the day. The only times you can actually find Mercury are when it appears farthest from the Sun (astronomers call this position of a planet its greatest elongation) while the Sun is still below the horizon. Sometimes you can catch Mercury just after sunset and other times just before sunrise. But without a telescope, you can never see Mercury in the middle of the day (the Sun is too bright) or late at night (because Mercury sets soon after the Sun and is below the horizon).

Here's where it gets complicated. Some elongations are better than others and some seasons make it easier to catch Mercury than others. Each year you get about three chances to see Mercury in the predawn sky and three chances to see it at twilight. The best months to see Mercury in the evening are April and May, and it can be easier to spot the elusive planet before sunrise in October and November. Your window to see Mercury at its greatest elongations (morning or evening) is small. You can only see Mercury far enough from the Sun for about a week around each elongation. Consult sky simulation apps or astronomy websites to find out when Mercury will be at its most favorable elongations for the year.

To see Mercury when it is at its greatest eastern elongation, you will want to look to the western sky 15 minutes after sunset. It will be low in the sky, so you will need a clear view to the western horizon free from buildings or trees. As the sky darkens search for a steady, semi-bright light glowing through the twilight. It will be about as bright as the brightest nighttime stars. But since it will be low in the sky, you might notice that it has a different color. Most observers describe Mercury as having a pinkish hue. The planet itself is actually dark gray, but the pink color shines for the same reason that a setting Sun turns from yellow to that memorable shade of red. The light bouncing off Mercury must travel through more layers of Earth's atmosphere to reach your eyes. This makes the light reflected off Mercury's surface appear a tinted pink.

To observe Mercury in the morning sky, you will have to wait for its greatest western elongation. When this occurs, look low above the eastern horizon 45 minutes before sunrise. Over the next 30 minutes you might see pinkish Mercury pop into view just before the Sun pokes above the horizon.

# 03

# MERCURY

**CLASSIFICATION: PLANET** | **VISIBILITY: EASY**

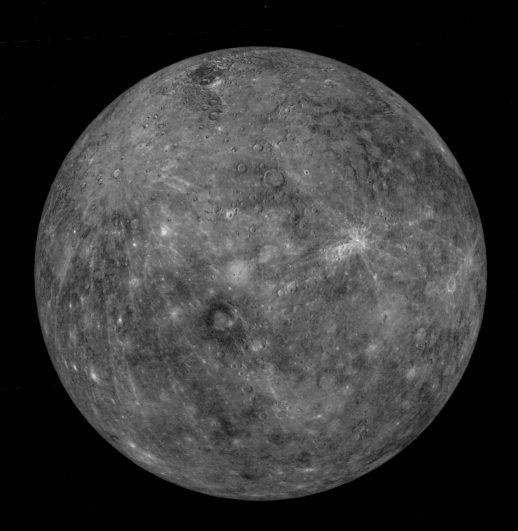

## SEE MERCURY THROUGH A TELESCOPE

Mercury's complicated relationship with the Sun makes this planet a challenge to observe clearly through a telescope. On those rare occasions when Mercury is visible and you can point a telescope toward it, your view may be disappointing. "Oh, that's it?" is a very common reaction to seeing the closest planet to the Sun through a telescope. First, Mercury is a very small planet—the smallest planet in the solar system—at only 3,032 miles in diameter (Earth's diameter is almost 8,000 miles). Plus, when you see Mercury it could be 50–100 million miles away. As a result, it will often look like a pinkish pinpoint of light even when you are viewing through a good telescope.

However, if you look closely you may notice that it is not a pinpoint or a perfect circle. Mercury can exhibit many shapes. At elongation Mercury should look like a tiny half-moon. And if you're really lucky and have a nice clear sky, it can sometimes appear as the little crescent. Why is this happening? Why does Mercury go through phases? Just like the Moon, Mercury gets all of its light from the Sun. You can observe different phases of Mercury depending on where it is in its orbit around the Sun. When it is at elongation, your perspective on the little planet only allows you to see half of it lit up. When it comes nearly between the Earth and Sun, Mercury looks like a little crescent.

## WITNESS A TRANSIT OF MERCURY

Since Mercury is the closest planet to orbit the Sun, does it ever come between the Sun and Earth? Can Mercury eclipse the Sun? Yes and no.

Yes, Mercury can pass directly between the Sun and Earth. But no, it is too small in our sky to completely eclipse the Sun. It will merely look like a tiny disc, a circular black blemish before the dazzling surface of the Sun. Astronomers call this a transit. Transits of Mercury are rarer than eclipses and occur only about 13 times per century. The next transits of Mercury will be November 13, 2032; November 7, 2039; May 7, 2049.

To observe a transit of Mercury, you will need to take the same precautions as you do when viewing the Sun. Protect your eyes by employing approved solar filters. And since Mercury seems so tiny in the sky, you will not be able to see this transit with the naked eye. You will need to magnify the image and outfit binoculars or telescopes with safe solar filters to see the alignment of the Sun and Mercury.

"The nitrogen in our DNA, the calcium in our teeth, the iron in our blood, the carbon in our apple pies were made in the interiors of collapsing stars. We are made of star stuff."

Carl Sagan, astronomer and author

# 04

# VENUS

| CLASSIFICATION: **PLANET** | VISIBILITY: **EASY** |

Venus was the goddess of beauty in ancient Rome and her brightly shining planetary namesake lives up to that hype. When Venus is in the evening sky, you notice her. When she is in the predawn sky, you can't miss her. When Venus shines, she looks like an extraterrestrial visitor has graced the heavens. There is something UFO-like about her.

As the Sun dips below the western horizon and twilight descends into a deeper shade of blue, one "star" bursts into view before any other. That "star" is the planet Venus. It is so incredibly bright in the sky because a thick blanket of clouds perpetually covers its surface. These clouds reflect so much sunlight into space that from Earth only the Sun and Moon shine brighter than Venus.

Venus is the second closest planet to the Sun. Like the other inner planet, Mercury, Venus can best be seen just before sunrise or just after sunset, depending on where it is in its orbit. Some people call Venus the Morning Star or Evening Star, depending on when it is visible. Venus looks especially good when it is near a waxing crescent Moon in the fading twilight of evening, or when it cozies up to a waning crescent Moon just before dawn.

## THE PHASES OF VENUS

Venus's brilliance comes at a price. The clouds that reflect so much sunlight completely obscure our view of the surface below. However, Venus presents a better sight through a telescope than Mercury. At 7,521 miles in diameter Venus is a much larger planet than Mercury, and it can come much closer to Earth. So even with a small telescope you can easily notice Venus's shape.

When Venus is on the other side of the solar system, it looks like a small disc. It's almost a perfect circle. But as Venus rounds the Sun and starts coming closer to Earth, it changes phases: first into a gibbous Venus (which is almost full), then into a half Venus, and then into a crescent Venus. Since Venus only shines from reflected sunlight, you can see different phases of Venus, depending on where it is in its orbit around the Sun.

Also, Venus's apparent size changes dramatically. At its farthest, Venus is about 162 million miles from Earth. But when you see it as a slim crescent, it is only about 40 million miles away. So Venus can look 4 times larger just before coming between us and the Sun. In fact, during the crescent-Venus stages, the planet is so close to us that some people can even discern its shape with the naked eye.

## SEE VENUS IN THE DAYTIME

If you're looking for a challenge, try finding Venus during the daytime. During the several months when Venus may be visible in the evening sky right after sunset, strive to spot it before the Sun sets. During greatest elongation, when Venus appears to be farthest from the Sun, this is definitely possible. In fact, if the western sky is really clear, during the greatest elongations you can find Venus 30 minutes before sunset.

A tougher challenge is to find Venus with the naked eye during the middle of the day. The bright sunlight washes out the light of every star, and so they are impossible to find during the daytime. However, Venus is bright enough that if you know exactly where to look, you can find it as a pale white dot amid the light blue midday sky. It is extremely difficult to locate Venus this way but the Moon can help.

The easiest times to find Venus in the daytime are during Moon-Venus conjunctions. These happen when the crescent Moon appears near Venus in the sky. Sky simulation software or phone apps can show you when and where the Moon and Venus will be in conjunction and how they'll appear (they reveal if Venus will be on the left or right of the Moon, for example). Then on the appropriate day, first find the Moon in the sky. Once you do that, slowly scan the sky around the Moon until you spot a little pale dot. If you see that dot, it has to be Venus, because other than the Sun and Moon you cannot make out any other planet or star in the daytime sky. It takes some practice and patience, but once you get the hang of it you will be able to find Venus in the daytime.

## VENUS'S 8-YEAR CYCLE

The ancient Mayan astronomers were obsessed with Venus. They charted its wanderings across the heavens with such precision that they could predict where Venus would be in the sky years in advance. From their countless observations, they discovered this pattern: Venus was visible in the morning sky just before sunrise for about 260 straight days. Then Venus went behind the Sun (from their perspective) and was not visible in the sky for about 56 days.

At the end of that period, Venus popped out into the evening sky and remained visible each night after sunset for another 260 days. But then something strange happened: When Venus came between Earth and the Sun, it was only invisible for a scant 8 days before reappearing again in the morning sky for another 260 days. They observed that the cycle repeated itself every 584 days.

The Maya also figured out that over a period of 8 years, this 584-day cycle repeated itself exactly five times. If you trace Venus's path in the sky at the same time every day, you'll notice that it makes a weird loop. And if you chart the course of Venus over 8 years, you'll record five distinct patterns: a squiggle down, a loop down, a zigzag, a loop up, and a squiggle up. That's five shapes that repeat almost exactly every 8 years.

The Mayan astronomers demonstrated their knowledge of Venus in a book of tables called the Dresden Codex. Their Venus tables were so accurate that after 500 years, their predicted positions of Venus would only be off by 1 day. You can watch Venus just as carefully as the Maya. Note what time of day you see it, how high in the sky it is, and where it rises or sets. Do that for 8 years and you'll have the pattern down pat!

"I remember on the trip home on Apollo 11, it suddenly struck me that that tiny pea, pretty and blue, was the Earth. I put up my thumb and shut one eye, and my thumb blotted out the planet Earth. I didn't feel like a giant. I felt very, very small."

Neil Armstrong, astronaut

Mars is a small planet that is only 4,212 miles wide (about half the diameter of Earth), but it is still extremely easy to find in the night sky with the naked eye. It is nick-named the Red Planet because it shines with an extremely off-white light. When you find it, you might classify its hue as orange or yellow instead of red, but when you compare its light to that of white or blue stars, you'll see it is redder than most. This bloody color led ancient Romans to associate this planet with Mars, their god of war.

In the sky Mars generally appears brighter than the brightest stars. Even when very far from the Earth, Mars is on par with the first magnitude stars (the brightest stars) that are visible from even urban locations. But when Mars is especially close to Earth it is a dazzling orange beacon in the night. When Mars is nearby, only Venus and Jupiter can shine brighter.

## MARS IN OPPOSITION

Mars is best viewed when it is closest to the Earth, and thus appears at its biggest and brightest in the evening sky. The Red Planet's distance from us varies significantly: from almost 250 million miles at its farthest to about 35 million miles at its closest. Astronomers use the term *opposition* to describe Mars's regular close passes to Earth. Opposition occurs when the Earth is between Mars and the Sun, placing the Sun on the opposite side of the sky from Mars. This friendly alignment happens about every 26 months.

Not all oppositions are created equal. Because Mars has an eccentric orbit, some oppositions are much closer than others. At the 2003 opposition Mars was about as close as it could possibly get to Earth: 34.5 million miles. Other oppositions place Mars at 45, 50, or even 60 million miles away. Each opposition provides a better opportunity to check out Mars more closely. Although the next exceptionally close opposition will be in September 2035, you don't have to wait that long to see more dramatic passes of Mars.

As the Sun sets in the west during opposition, you can spy the Red Planet rising in the east and blazing in the evening sky with a steady ruddy glow. Mars's dates of opposition and the constellation in which you can find it from 2020–2030 are:

- October 13, 2020, in Pisces
- December 8, 2022, in Taurus
- January 16, 2025, in Gemini
- February 19, 2027, in Leo
- March 25, 2029, in Virgo

If you miss seeing Mars on those dates, don't worry. Those are just technically the best dates to observe Mars. Mars will still be almost as bright for about a month before and a month after opposition.

After opposition, Mars tends to hang around in the evening sky for a long time. You will see it drifting slowly to the west, night after night, for 6 to 9 months after opposition. It barely changes positions from night to night. This happens because Earth and Mars are traveling around the Sun in the same counterclockwise direction. Because Earth is on the inside track, it is moving faster than Mars, and so each night Mars drifts farther and farther away and appears dimmer and dimmer in the night sky.

| # MARS

CLASSIFICATION: **PLANET** | VISIBILITY: **EASY**

## THE PECULIAR MOTION OF MARS

Like all planets, Mars wanders across the night sky, appearing to travel from constellation to constellation. However, it does not keep up a steady pace. It goes along, slows down, stops, reverses itself, slows down again, and then goes forward once more. This pattern repeats every 26 months. If you took a picture of Mars every night at the same time, you would see that it makes a loop-de-loop pattern. So what is going on here? This is Mars's retrograde motion, and although this distinctive motion was well known to ancient astronomers thousands of years ago, it took until the 1500s for a Polish astronomer to correctly explain it.

The ancient models of the universe placed an unmoving Earth at the center surrounded by the circular pathways of the Sun, Moon, and planets (Mercury, Venus, Mars, Jupiter, and Saturn). But if the planets just circled Earth, why did Mars make a loop-de-loop exactly when the Sun was on the other side of Earth? Ancient astronomers scrambled for answers, and they came up with a mechanism called an epicycle. Basically, they proposed that Mars completed a circle upon a circle. Eventually the astronomers needed epicycles upon epicycles to make their models match what appeared in the sky above. It became really, really complicated, but it worked for over a millennium.

When the Polish mathematician and astronomer Nicolaus Copernicus checked on the planets in the early 1500s, he discovered that Mars and Saturn weren't where they were supposed to be. He thought that the epicycles were too complicated and obviously flawed, so Copernicus began searching for an alternative explanation. After intently studying the heavens, he explored what would happen if the Sun was at the center of the universe instead of Earth. Copernicus then hypothesized that Earth and Mars both travel around the Sun, but Earth moves faster. As it catches up to Mars and passes it, the Red Planet appears to move backward in the sky. The loop-de-loop was just part of a grand illusion. Earth actually moved! Copernicus's Sun-centered universe was perhaps one of the greatest discoveries of all time, and it was the motion of Mars that provided the biggest clue.

## MARS IN A TELESCOPE

You can always find Mars near one of the twelve zodiac constellations (i.e., Leo and Pisces). Many star charts and astronomy apps will tell you in which constellation to look for Mars and what time of night it is easiest to see. Before it reaches opposition, you can find Mars in the predawn sky in the south. During opposition it rises in the east just after sunset. And after opposition Mars hangs out in the southern sky after dark. No matter the season, if Mars is in the sky it will be brighter than almost every star in the surrounding constellation in which it resides.

When you find Mars in a telescope, do not expect to see anything nearly exciting as Martians. Although Mars may be the most fascinating planet, through a backyard telescope it can appear underwhelming. Mars is a small planet and it looks like a small orange disc through most telescopes. However, when Mars is closest to Earth you may be able to detect features on the Martian surface.

If the planet is properly tilted, you may be able to discern a white spot marking a polar ice cap. At other latitudes on the surface you may see darker markings such as lines, blobs, and triangles. These were known as the Martian canals in the nineteenth and early twentieth centuries. We now know that these so-called canals are merely darker colored rocks that show up against the rusty soil of Mars.

The most dramatic Martian feature you can see through a telescope is called Syrtis Major. It looks like a dark brown triangle or chevron on the orange surface of Mars, but it is in fact the remnants of a shield volcano that has been dormant for eons.

Whether you observe Mars with the naked eye, binoculars, or a telescope, it may inspire you to dream of Martians visiting Earth someday or of Earthlings crossing the millions of miles of space to colonize a new planet. Let your imagination go wild under the glow of the Red Planet.

# 06 | JUPITER

CLASSIFICATION: **PLANET**  VISIBILITY: **EASY**

Ancients Greeks called it Zeus, and the Romans adopted this bright night light as the manifestation of their chief god, Jupiter. How did the ancients know Jupiter was the king of the planets? It's not the brightest planet—that's Venus. It's not the fastest- or slowest-moving planet—that's Mercury and Saturn, respectively. There must be something kingly about Jupiter that was visible even to the naked eye.

Jupiter is by far the largest planet in our solar system. It has an equatorial diameter of about 88,000 miles, meaning that 1,325 Earths could fit inside it. In fact, Jupiter is more massive than all of the planets, moons, and asteroids in the solar system combined! So its name really fits what we know about this giant planet.

### IDENTIFYING JUPITER

Jupiter is an unmistakable light in the night sky. It appears to be a non-twinkling cream-colored star and is very often the brightest starlike object in the entire night sky. Jupiter's brilliance is so stunning that it is often more than twice as bright as the brightest star in the sky, Sirius. When Jupiter is in the sky, you notice it!

Jupiter is a steady performer. Unlike Mercury or Mars, the light it shines on Earth does not fluctuate very noticeably in brightness over the time it is visible in the night sky. Even though Jupiter is far from Earth (roughly 400 million miles), it is so large and its cloud tops reflect so much sunlight that it shines brightly whenever it is up in the sky. Even when it is on the other side of the solar system and slightly farther from Earth, Jupiter dazzles.

Like all planets, you can always find Jupiter hanging out among the stars of the zodiac constellations. It wanders among the constellations much more slowly than any planet we have studied so far. In fact, it takes Jupiter 11.86 Earth years to orbit the Sun. That means each year Jupiter shifts its position in the sky by one-twelfth of a circle. This circle is called the zodiac, and Jupiter visits each zodiac constellation in order, one by one, year after year. For instance, if you see Jupiter in front of the stars of Libra, the next year it will be in front of Scorpius. The year after that it will grace the presence of Sagittarius, and so on. About every 12 years Jupiter will return to nearly the same location in the sky, completing a grand tour of the zodiac. Check astronomy websites and star charts to know in which zodiac constellation to look for Jupiter this year.

### JUPITER AT ITS BRIGHTEST

Like Mars, Jupiter is closest to Earth near its opposition. Jupiter takes almost 12 years to revolve around the Sun, and so Earth catches up to it about every 13 months. Thus, oppositions, and the best season for Jupiter-viewing, occur about 13 months apart.

The oppositions of Jupiter between 2020 and 2030 are listed here. You don't have to look for Jupiter only on these dates, however. You can find the giant planet shining brightly for several months before and after opposition. Astronomers just like to be technical and give you the absolute best nights to observe Jupiter and where to look, which are:

- July 14, 2020, in Sagittarius
- August 20, 2021, between Capricornus and Aquarius
- September 26, 2022, in Pisces
- November 3, 2023, in Aries
- December 7, 2024, in Taurus

- January 10, 2026, in Gemini
- February 11, 2027, between Cancer and Leo
- March 12, 2028, in Leo
- April 12, 2029, in Virgo
- May 13, 2030, in Libra

## JUPITER THROUGH BINOCULARS

If you view Jupiter through a pair of binoculars, you may be in for a special treat. Although you may not be able to see any features on the planet itself, you should be able to pick out a few little dots next to the planet. These are Jupiter's largest moons: Io, Europa, Ganymede, and Callisto. There are some rare humans who possess seemingly superhuman eyesight and can see these four moons with their naked eyes. Test your eyesight and see if you can do it too. If not, get some binoculars—or, better yet, a telescope.

## JUPITER THROUGH A TELESCOPE

With even a small telescope, Jupiter looks amazing. You can re-create the observations of Italian astronomer Galileo Galilei, who turned his homemade telescope toward Jupiter in 1610. Galileo saw Jupiter's four largest moons (now called the Galilean Moons) and computed their orbits. Each night Io, Europa, Ganymede, and Callisto form a different pattern. One night there are two on one side and two on the other. Another night you might see three on one side and only one on the other. And sometimes you'll only see three moons because the other is hiding in front of or behind the planet.

Observe Jupiter and its moons every night for a week or two and see if you can figure out the pattern to their motions. How can you even tell which is which? Somehow, Galileo could tell them apart more than 400 years ago.

Through a midsize or a large telescope Jupiter really comes to life. You can see the disc of the planet, two major dark bands, several smaller bands, and, sometimes, the Great Red Spot. This persistent mega-hurricane on Jupiter is always located in one of the thicker dark bands that ring the planet. You may first note the Great Red Spot's presence as a break in one of the bands, like a chunk is missing from it. But as you look more closely, you may be able to see this circular spot that is in reality much larger than Earth.

When one of the Galilean Moons comes between the Sun and Jupiter, you can tell. The moon will cast a shadow onto the surface of the planet. It looks like a little black spot on Jupiter. Occasionally you can catch two or even three circular shadows on Jupiter at one time. As you can see, Jupiter truly is one of our most remarkable planets.

"In one of those stars I shall be living. In one of them I shall be laughing. And so it will be as if all the stars were laughing, when you look at the sky at night."

Antoine de Saint-Exupéry, writer and poet

As "wandering stars" go, Saturn is the slowest-moving planet. Whereas Mercury whips around the Sun at over 100,000 miles per hour, Saturn practically pokes along at about 22,000 miles per hour. That means that although Saturn wanders across the background stars like its fellow planets, it does so at an incredibly slow pace.

It takes weeks, months, or even years to note any change in Saturn's position relative to the background constellations. While Earth circles the Sun every year, it takes Saturn about 29.5 years to complete an orbit. If you observe Saturn one starry night and come back to it a year later, the stars will be in the same place, but Saturn will have moved only about 12 degrees to the east. That means if you see Saturn among the stars of Sagittarius it can take 2 to 3 years for Saturn to move on to the next constellation of Capricornus.

The ancient Greeks and Romans noticed this sluggish motion and incorporated it into their mythology. Saturn (also known as Cronus in Greece) was the father of Jupiter (Zeus). After Jupiter overthrew his father and became the supreme god, Saturn became an old man and receded into the background. Saturn is often depicted as an old man with a long beard and later was equated with the figure of Father Time.

### IDENTIFYING SATURN

The farthest planet you can see with the naked eye, from even suburban locations, is Saturn. Although you cannot detect the ridiculously cool rings of this planet without a telescope, you can still easily locate it every year in the night sky.

Saturn appears to be a non-twinkling yellow star that shines with a light equal to or sometimes greater than the brightest first magnitude stars like Vega and Arcturus. Saturn definitely does not stand out as much as Venus, Mars, or Jupiter, but once you find it you will be able to go back to it night after night since it changes positions so slowly. Saturn never strays far from the zodiac constellations, so you can check astronomy websites or sky simulation apps to discover what area of the sky the ringed planet will be in for more than a year in advance.

### SATURN AT ITS BRIGHTEST

Like all planets farther from the Sun than Earth, Saturn appears biggest and brightest near opposition. At opposition Saturn is still over 800 million miles away, but it is so large and its surface and rings reflect so much sunlight that it shines like a first magnitude star.

Since Saturn is so slow-moving, Earth can round the Sun and catch up to it almost every year. The following list contains the upcoming opposition dates for Saturn and in which constellation you can find it. Notice that there is only an 11- or 12-day difference in dates from year to year. Just like Jupiter, you can see Saturn shining nearly as brightly up to 1 month before and 1 month after opposition. The dates are:

- July 20, 2020, in Sagittarius
- August 2, 2021, in Capricornus
- August 14, 2022, in Capricornus
- August 27, 2023, in Aquarius
- September 8, 2024, in Aquarius
- September 21, 2025, in Pisces
- October 4, 2026, in Pisces

# SATURN

CLASSIFICATION: **PLANET**     VISIBILITY: **EASY**

- October 18, 2027, in Pisces
- October 30, 2028, in Aries
- November 13, 2029, in Taurus
- November 27, 2030, in Taurus

### SATURN SHINES IN A TELESCOPE

What are the best things to see in a telescope? Saturn, Saturn, and Saturn.

It is a rite of passage to locate Saturn and observe it in a telescope. When you swing a telescope toward Saturn, center it, and place your eye to the eyepiece, something magical will happen. There Saturn will sit among the blackness of space, a tiny cartoon world encircled by rings. It will look totally fake—as if someone placed a sticker of Saturn at the end of your telescope. But it is the real thing! You are experiencing sunlight bouncing off an improbable planet almost 1 billion miles away, coming through the telescope and into your eye.

Saturn has a transformative power that excites our imagination like no other astronomical object. Saturn is our childhood symbol for space, and seeing the real thing literally lights up our faces.

Most telescopes can allow you to detect several of Saturn's moons. Titan is by far the brightest and will look like a steady-shining star off to one side of the planet. The moons Rhea, Tethys, and Dione can be spotted with a moderate telescope. If you're lucky enough to view Saturn under ideal conditions through a large telescope, you may be able to make out the fainter moons Iapetus, Enceladus, and Mimas as well.

### RING TILTS AND SATURN SEASONS

Every year Saturn looks a little different in a telescope. During its 29.5-year orbit of the Sun it's tilted by 27 degrees, with respect to its orbit around the Sun. This slant changes your viewing perspective on the rings. When the rings are tipped up or down to you, Saturn reveals the classic views you would expect. When they are tilted to the max, you can behold the breadth of the rings, observe their structure and gaps, and witness how they cast shadows onto the planet.

The rings are surprisingly svelte, and when they close down and become edge-on to you, they become incredibly difficult to see. In some places Saturn's rings are barely 100 feet deep. That means when the plane of the rings is pointing directly at Earth, they are invisible through even the most powerful telescopes. When Saturn seems to be missing its rings, astronomers sometimes call this "naked" Saturn. The next naked Saturn will occur in 2025, but until then make it your goal to see Saturn in a telescope. You will not be disappointed.

"The Sun, with all those planets revolving around it and dependent on it, can still ripen a bunch of grapes as if it had nothing else in the universe to do."

Galileo Galilei, astronomer and physicist

# Part II.

# Stars and Constellations

Ancient stargazers marveled at all of the lights above—the different luminosities, shades, shapes, and bands of stars. The stars were gods, unreachable but ever-present with hints of personality. Stargazers watched how the stars rose, crossed the sky, and set. Some stars even fluctuated in brightness and brought intrigue to the masses. Ancient stargazers from Nigeria to Egypt, China to India, the Americas, and Europe connected the dots in the sky and formed pictures called constellations.

You are now going to take up the pastime of your ancestors. You will hear the stories and look for the same stars they observed thousands of years ago. The positions and brightness of the vast majority of stars have not changed significantly in the last four millennia, and as a result the outlines of the constellations have not drastically altered. So you will essentially be seeing the same sky with the same configurations of heavenly lights that have glittered and glistened on every great civilization throughout human history.

In this part I've broken the sky down into five sections. First, we will explore the northern sky and stars that are visible almost all year round from the mid–Northern Hemisphere. Then we will take a closer look at the stars and constellations visible by season: the winter sky, spring sky, summer sky, and fall sky. Each section will start with a wide-angle view of the firmament in order to get a better overview. Then we will zoom in on individual constellations and visit the most interesting stars that reside within them. Since timing is important here, please note that the charts display the view of the sky during evening hours—prime-time viewing for most people. Let's head to the sky!

# The Northern Sky

For sky watchers living in the Northern Hemisphere, the stars in the northern sky hold a special place of honor. Over the course of every night these stars rotate around a central point. After an entire day each northern star makes a complete circle and returns to its starting point.

These stars are called circumpolar stars, meaning they circle the celestial pole and never set. They are visible every night of the year. From the midlatitudes (like the United States, Europe, and much of Asia) the circumpolar constellations include Ursa Major; Ursa Minor; Cassiopeia, the Queen; Cepheus, the King; and Draco, the Dragon. Although their elevations may change in the northern sky, with a little practice you can find them easily in relation to the Big Dipper.

Picture the stars and constellations in the northern sky as if they were etched into a circular plate. At the center of the plate lies the most famous and important star in the sky, Polaris, also known as the North Star. The stars of the Big Dipper are on one side of Polaris, and Cepheus and Cassiopeia are on the other side. Draco's tail begins between the two Dippers and its body coils around the Little Dipper.

These shapes seem fixed onto this plate, and once a day the plate spins counterclockwise around Polaris. So when the Big Dipper rises on one side, Cassiopeia goes down on the other. Draco's head can rise very high in the sky above the Little Dipper, and 12 hours later circle very low in the sky below it.

When you watch the northern sky for an hour or two, you can very easily detect this counterclockwise motion. If you set up your camera on a tripod and take a very long exposure, you can capture star trails—arcs traced by the light of the stars moving around Polaris. This grand circle is caused by the rotation of Earth, the same motion that makes the Sun rise and set. This celestial circumnavigation is marvelous to experience under a dark sky. Let's take a look at these northern sky stars and constellations.

53

# Northern Sky Constellations

Alcor

Mizar

**URSA MAJOR**

Dubhe

Merak

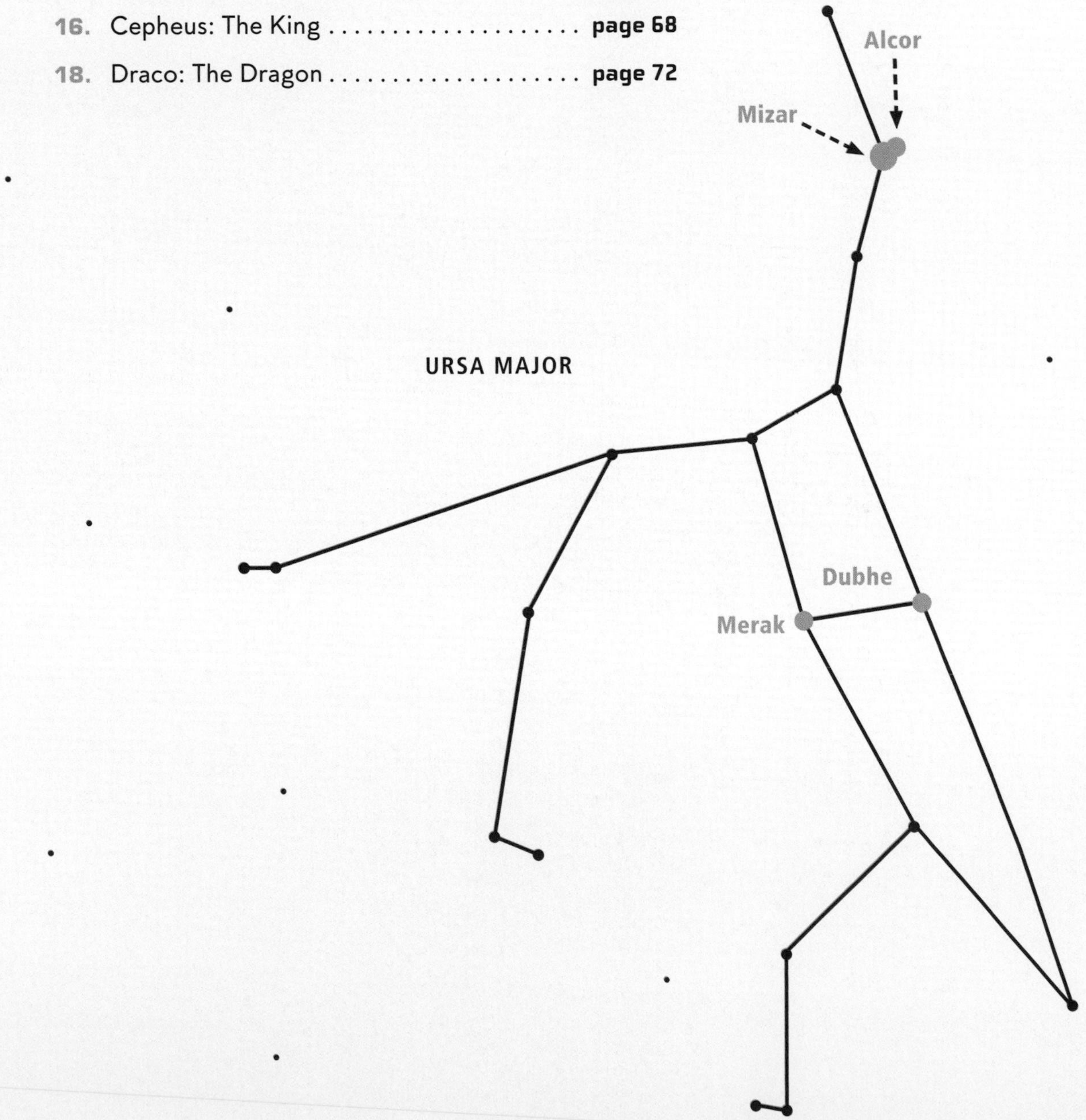

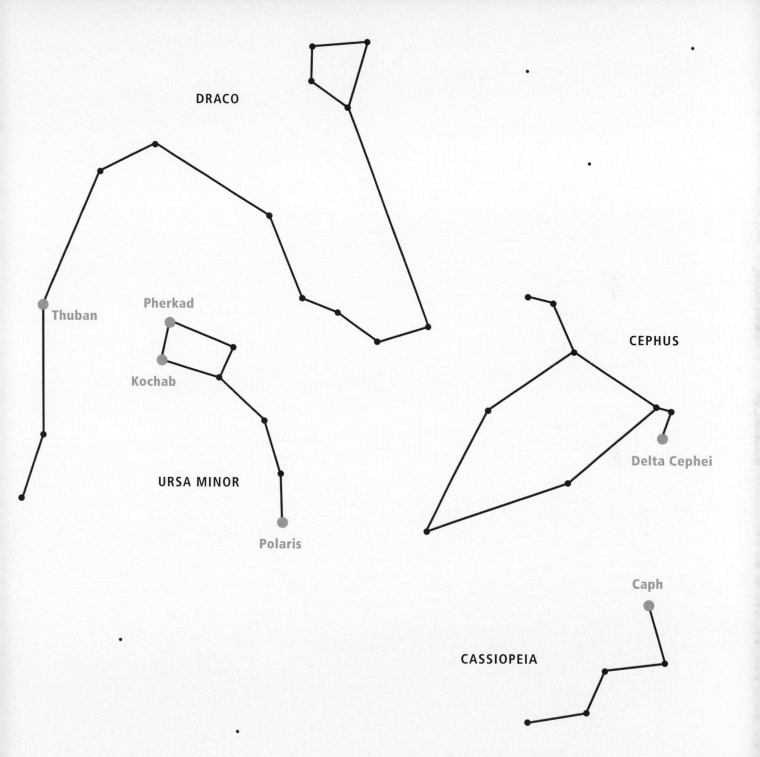

DRACO

Thuban

Pherkad

Kochab

URSA MINOR

Polaris

CEPHUS

Delta Cephei

Caph

CASSIOPEIA

# 08

# URSA MAJOR

## *The Big Bear*

| CLASSIFICATION: **CONSTELLATION** | VISIBILITY: **EASY** |
| --- | --- |

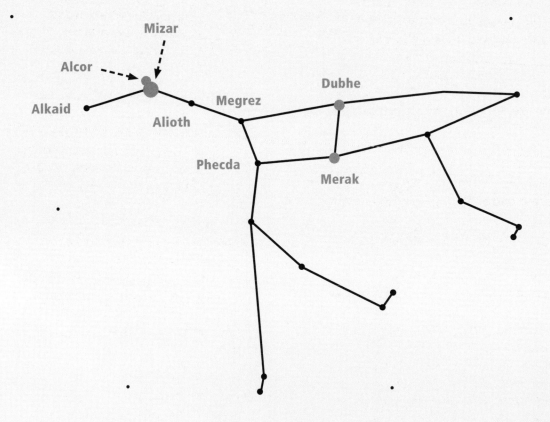

Ursa Major, a constellation that both the ancient Greeks and some Native American groups identified, shows a bear with a very long tail. As the Native American legend goes, once upon a time a hunter came across a momma bear blocking his path. To move her out of the way, the hunter grabbed her by her short, stubby tail and began swinging her over his head. And as the bear was twirling around and around, a funny thing began to happen. The bear's tail started stretching and stretching until finally the hunter let go. The bear flew up, up, up so high that she stuck into the sky where she slowly turned into the stars we see today. And that is how the Big Bear got her long tail.

The most recognizable star pattern within Ursa Major is the Big Dipper, a major landmark of the northern sky. Four stars form the bowl of the spoon and three more make a curved handle. Many people think the Big Dipper is a constellation, but it's actually an asterism, a recognizable shape of stars. The Big Dipper forms the rear end and long tail of the Big Bear. Fainter stars nearby suggest the big bear's legs, paws, and head in this sprawling structure.

Other cultures looked at the shapes of these Big Dipper stars and created much different sky pictures. Fishing cultures called them the fish hook. In England it was the plow. In medieval Europe it was a stretcher, wagon, or cart used to carry dead bodies. And among African Americans it was called the Drinking Gourd. With your modern imagination you can see it as a shopping cart or a lawn mower. And sometimes the Big Dipper can look like a question mark when it is standing on its handle.

## HOW TO FIND IT

**1** To find Ursa Major in the night sky, just look up and find the familiar pattern of the Big Dipper. Note that the Big Dipper will be much more prominent in the sky than its smaller counterpart the Little Dipper. Both Dippers lie in the northern sky and are only about 25 degrees apart. If you live in or near city lights, chances are you will not be able to see the entire Little Dipper. So if you see a Dipper from an urban location, it is the Big Dipper. If you are stargazing out in the countryside and you can see the Little Dipper in the sky, the Big Dipper will also be visible and will be so much more prominent.

**2** Now, although Ursa Major circles the North Star once a day, its position in the sky changes from season to season. Every **WINTER** evening you can find the Big Dipper standing on its handle in the northeastern sky. During the **SPRING** it rides high in the north above Polaris with its cup turned down. On **SUMMER** evenings the Big Dipper stands on its spoon in the northwestern sky. The **FALL** is often the toughest time to see the Big Dipper because it scoops low in the sky, just above the northern horizon, and it may be difficult to locate above the treetops.

**3** No matter where the Big Dipper is, the stars that construct the entire constellation Ursa Major will be around it in the same formation.

# 09

# MERAK AND DUBHE

## *The Ultimate Pointer Stars*

| CLASSIFICATION: **STARS** | VISIBILITY: **EASY** |

Merak and Dubhe are pointer stars: prominent, recognizable stars that point to less conspicuous stars in the sky. These are perhaps the best examples of pointer stars because they will show you the way to the North Star and the constellations Cepheus, the King; Cassiopeia, the Queen; Andromeda, the Princess; and Pegasus, the Flying Horse.

Within the constellation Ursa Major many of the stars that form the Big Dipper are about the same distance from Earth—about 80 light-years away. They most likely formed from the same nebula a long time ago and have drifted apart. Merak is one of the stars in this group. It should look like a second magnitude white star but is actually about 63 times more luminous than our Sun. If it was closer to us, it would be exceptionally bright. The name Merak comes from the Arabic word meaning "loins." So now you know what part of the bear Merak is supposed to represent.

Dubhe is much farther away than Merak and did not form with the others. It only looks to be part of the group. Dubhe is the farthest Big Dipper star at 123 light-years from Earth and shines with a yellowish tint. Its name means, simply, "bear" in Arabic.

## HOW TO FIND IT

1 Merak and Dubhe are the two stars on the end of the bowl of the spoon of the Big Dipper.

2 These stars can be found in the northern sky almost all year round. Only in the **FALL**, for a brief time, are they too low to see above the horizon. But over the course of the night they will rise higher and continue to circle the North Star.

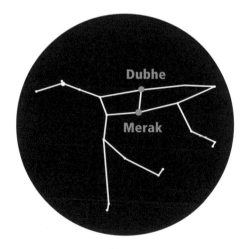

# 10 | MIZAR AND ALCOR

## and the Ancient Eye Exam

CLASSIFICATION: **STARS** | VISIBILITY: **MODERATE**

Mizar is one of the seven stars in the Big Dipper. Upon first glance it may only look like a single star, but if you have good eyesight you may be able to detect a dimmer companion star right next to it called Alcor. But there is even more to Mizar and Alcor than meets the eye. They are the brightest pair in a sextuple system of stars (six stars that circle around each other). The four other stars in the system cannot be seen without a telescope.

*Mizar* means "the apron" or "the wrapper," and fainter Alcor was referred to as "the forgotten" or "neglected one." In Europe the pair was often associated with horses and was known as the horse and rider—the brighter star Mizar being the horse, and the dimmer Alcor being the rider. In Indian astronomy the stars in the Big Dipper were the Saptarishi, the Seven Sages. Mizar was known as Vashistha, and Alcor represented his wife, Arundhati. The proximity of these two stars in the sky symbolized the unity of marriage.

Mizar and Alcor had a special significance to ancient stargazers from the Americas to the Mediterranean. Legend has it that they used a star from the Big Dipper as a kind of eye exam. If you wanted to be a scout or hunter, you had to prove that you had the eyesight for the job. An elder would take prospective scouts out under the stars, point toward the Big Dipper, and ask which of the seven stars may be more than met the eye. If the job applicant had nearly twenty-twenty vision, he would notice that Mizar was not alone in the sky but had a small, faint companion right next to it, the star called Alcor.

### HOW TO FIND IT

**1** Three stars create the Big Dipper's handle and Mizar is the middle star of these three. It is a second magnitude star and is easy to locate.

**2** Alcor is about one-quarter as bright as Mizar so it can be much more challenging to spot. Squint and look carefully near Mizar and you may just make it out. Sometimes Alcor even looks like a small bump on Mizar.

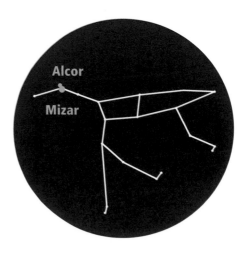

# URSA MINOR

## *The Little Bear*

| CLASSIFICATION: **CONSTELLATION** | VISIBILITY: **MODERATE** |

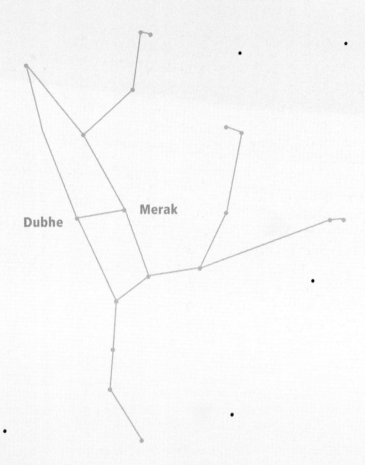

URSA MAJOR

Dubhe

Merak

URSA MINOR

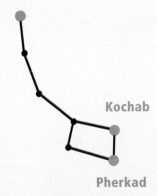

Polaris

Kochab

Pherkad

The Native American legend that describes the origin of Ursa Minor (or the Little Bear) can be considered the sequel to the Big Bear story. After the hunter threw the momma bear up into the sky (becoming the constellation Ursa Major), he continued on down the path. Soon he came across the saddest, most pathetic looking baby bear that had lost his mother. The hunter felt bad about separating the two. After all, he had just thrown the momma bear up in the sky a few minutes ago, and there she was shining down on everyone. But the hunter couldn't climb up into the sky to bring momma back, so he could only think of one solution: He grabbed the baby bear by his short, stubby tail and swung him around over his head. And soon the Baby Bear's tail began stretching and stretching until the hunter let go. The baby bear flew up and stuck in the sky not far from his momma.

Just like Ursa Major houses the Big Dipper, Ursa Minor houses the Little Dipper, another asterism. The four stars in the Little Dipper's cup form the little bear's body, while the three stars in the handle make an astonishingly long and curved tail. To see the entire outline of the little bear, you will need a lot of imagination and be far from city lights. From most light-polluted areas, you may only be able to see one, two, or three of the seven stars in this grouping.

## HOW TO FIND IT

**1** The easiest way to find Ursa Minor is to find its brightest star, Polaris (a.k.a. the North Star). First locate the Big Dipper and connect the stars on the end of the spoon's bowl, Merak and Dubhe. Continue that line straight away from Dubhe and you will run right into the North Star, which doubles as the end of the Little Dipper's handle.

**2** Since Ursa Minor is a circumpolar constellation, it is visible all year. Moreover, since it houses the North Star, you can find it every night in the northern sky. At the start of **SPRING**, in early evening, the Little Dipper resembles a ladle as if its bowl was turned properly to hold water. In the **SUMMER** it appears to be standing on its handle with the bowl higher in the sky. In the **FALL** its outline seems tipped over and can give the impression that it is emptying its contents. And in **WINTER** the Little Dipper stands on its bowl with the handle resting higher in the sky.

**3** No matter the time or season the North Star is always at the end of the spoon's handle and is always almost due north.

## 12 | POLARIS

### *The North Star*

CLASSIFICATION: **STAR** | VISIBILITY: **EASY**

Polaris, the North Star, is a white star about 433 light-years from Earth. That means whenever you see Polaris shining in the northern sky, you are really seeing the light that left that star 433 years ago. Many people think the North Star is the brightest star in the sky, but it actually ranks about forty-seventh in brightness of all the stars you can see from Earth.

What makes Polaris so special is that, by pure coincidence, Earth's North Pole points almost directly toward this star. As Earth spins, the North Star appears to remain fixed. From the Northern Hemisphere, Polaris shines due north in the sky all night, every night. Once you locate this star, you can establish your directions: north, south, east, and west. And if you watch it night after night, it may give you the impression that a bright nail has been hammered into the sky around which the dome of heaven slowly spins.

#### HOW TO FIND IT

**1** To find this star, connect the dots on the pointer stars Merak and Dubhe in the Big Dipper. Continue that line away from Dubhe and star hop over to the North Star.

**2** The distance between Dubhe and Polaris is about 28 degrees: the size of your open hand—the space between the end of your thumb and the tip of your pinkie—at arm's length, or the equivalent of five Merak-Dubhe distances.

**3** The North Star's angle above the horizon in the Northern Hemisphere is also equal to your latitude on Earth. For instance, if you live in New York City, Polaris will be 40 degrees up. In Miami it will be 25 degrees above the northern horizon. If you were on the North Pole, the North Star would be straight above you. And from the Southern Hemisphere, you cannot see the North Star at all. Knowing your latitude can help you narrow down how high or low in the sky to look for the North Star.

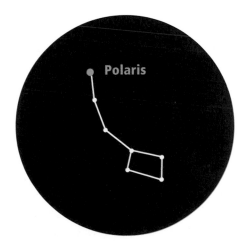

# 13 | KOCHAB AND PHERKAD

## The Guardians of the Pole

| CLASSIFICATION: **STARS** | VISIBILITY: **MODERATE** |

Almost 4,000 years ago the North Pole of Earth pointed between two stars within Ursa Major, called Kochab and Pherkad. Around 1900 B.C., Kochab actually served as the North Star. It helped guide travelers for hundreds of years while Polaris was just another ordinary star nearby. What changed in 4,000 years that turned Polaris into the North Star?

Think of Earth as an imperfectly spinning top. In addition to Earth's daily rotation upon its axis and yearly revolution around the Sun, it also has a slow wobble, which is called precession. This wobble is so slow that it takes 26,000 years to complete one cycle. Precession actually aims Earth's North Pole in a different direction. So after 4,000 years (and one-sixth of a wobble) the North Star changed from Kochab to Polaris. Which star will be the next North Star? You will find that out when you read about the constellation Cepheus later in this section.

Together, Kochab and Pherkad were known to ancient Egyptian astronomers as The Indestructibles. They are only 16 and 18 degrees away, respectively, in the sky from Polaris. This close proximity to Polaris has earned them the nickname "Guardians of the Pole."

In addition to Polaris, Kochab, and Pherkad, four other stars make up Ursa Minor (or the Little Dipper). These stars are too faint to see from urban locations. But maybe when you get under a truly dark sky, you can see the little bear in all its glory: body and stretched-out tail.

## HOW TO FIND IT

**1** To find these stars look no further than the Little Dipper. Kochab is the brighter star on the end of the bowl of the spoon of the Little Dipper. An orange giant about 130 light-years from Earth, Kochab is a second magnitude star and is about the same brightness as the North Star.

**2** Pherkad is the dimmer star on the bowl of the spoon and is about 487 light-years away from Earth. Kochab looks to be brighter than Pherkad only because of its proximity. If you placed the two stars and the Sun at equal distances from Earth, Kochab would be 390 times more luminous than the Sun, while Pherkad would be about 1,100 times brighter than the Sun.

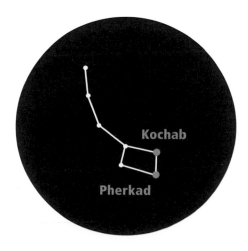

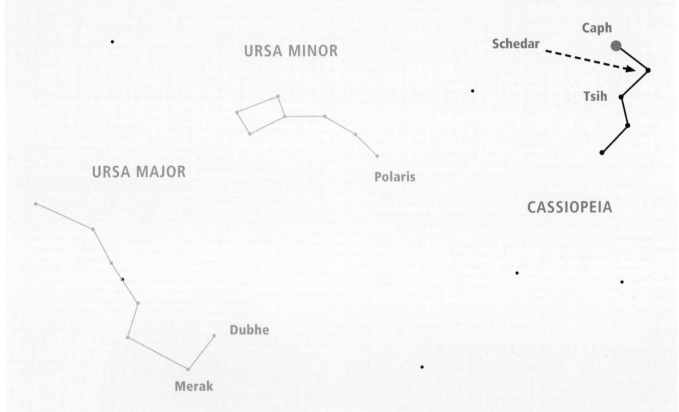

## 14 | CASSIOPEIA

### *The Queen*

| CLASSIFICATION: **CONSTELLATION** | VISIBILITY: **EASY** |

According to Greek mythology, Cassiopeia was the queen of Ethiopia who had a narcissistic habit. She believed that she was the most beautiful woman in the whole world, and she wasn't shy about telling people about her exquisiteness. One day, the queen went too far and proclaimed that she was more beautiful than all the sea nymphs. When the sea nymphs complained to their father, the god of the sea, Poseidon, Cassiopeia had to pay a terrible price for her vanity. In one version of the story Poseidon condemned her to the stars and turned Cassiopeia into a constellation. Since she landed near the North Star, she was fated to turn end over end, night after night. Some hours she sat comfortably on her throne, but most of the time she was sideways or upside down, hanging on to that throne for dear life. This was her ultimate punishment, because how could she be beautiful and upside down at the same time?

This myth aside, you may be surprised to learn that the stars that make up the constellation Cassiopeia do not look like a queen sitting on her throne. In fact, Cassiopeia is notorious for discouraging would-be stargazers because her outline is so ridiculously un-queen-like. That said, her five stars are semi-bright and make a very recognizable pattern that is visible almost all year. Instead of looking for a queen, merely find a small group of five stars that resemble a squished letter M, or W. If you still want to use your imagination, the stars in Cassiopeia could make a star-studded crown for our starry queen.

## HOW TO FIND IT

**1** To find Cassiopeia, first find the Big Dipper. Connect the line between the stars Merak and Dubhe. Continue that line of sight away from Dubhe and go to Polaris. Once you reach Polaris, keep going on that same line and hop over to the star named Caph (which you'll learn about in the following entry) at one end of Cassiopeia. From Caph you can then identify the M-shape of stars and know you've found the vain queen.

**2** Since Polaris is between Cassiopeia and the Big Dipper, when one star pattern gets lower in the sky, the other ascends. During the **WINTER** months Cassiopeia sits high above the North Star in the evening, sporting her classic M-shape. On **SPRING** evenings she is tipped over and can be found lower in the northwestern sky. Right after dark in the **SUMMER** Cassiopeia is just above the northern horizon and is a W-shape of stars. In the **FALL** she resembles a crudely drawn number 3 and sits higher in the northeast.

# 15

# CAPH

## *The Circler*

| CLASSIFICATION: **STAR** | VISIBILITY: **EASY** |
| --- | --- |

Caph is a white giant star about 54 light-years from Earth. Its name is derived from an Arabic word meaning "hand" or "palm." In the first millennium A.D. astronomers in the Middle East considered this pattern of stars to be a hand stained with henna, the five main stars forming the fingers.

Keep in mind that Caph is not the brightest star in Cassiopeia. It is just the easiest one to find and can point you to several others within the constellation. The honor of brightest star in Cassiopeia is shared between nearby Schedar and the star at the central peak of the M, named Tsih. Sometimes you can spy the triangle formed by Caph, Schedar, and Tsih before seeing the full M-shape of five stars. The fourth and fifth stars (those farthest from Caph), named Ruchbah and Segin respectively, can be tough to detect from light-polluted skies.

### HOW TO FIND IT

**1** Caph is located on the brighter left side of the M shape of Cassiopeia. It is almost a straight line of sight from Merak and Dubhe to Polaris and on to Caph. Each star in the line is almost equally bright, and the distance between Dubhe and Polaris is equal to the distance between Polaris and Caph.

**2** During the **WINTER** months, when Cassiopeia is high in the sky, Caph is on her left side. During the **SPRING** this star is at the bottom of the five-starred figure. During the **SUMMER** months Caph is on the right side of the constellation, and in the **FALL** it is at the top of the star pattern.

**3** Caph leads the way in a nightly spin for Cassiopeia. It is the leading star that seems to drag the entire constellation on a counterclockwise journey around the North Star. Night after night, year after year, Caph circles the Pole. Every 24 hours it completes a 361-degree spin, a little more than one complete circle.

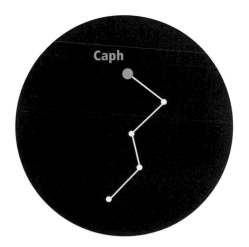

Caph

"Here men from the planet Earth first set foot upon the Moon July 1969, A.D. We came in peace for all mankind."

Apollo 11 plaque left on the Moon

# 16

# CEPHEUS

## *The King*

| CLASSIFICATION: **CONSTELLATION** | VISIBILITY: **DIFFICULT** |

DRACO

CEPHEUS

URSA MAJOR

Delta Cephei

URSA MINOR

Dubhe

Merak

Polaris

Errai

Caph

Segin

CASSIOPEIA

In Greek mythology Cepheus was the king of Ethiopia and husband to the queen, Cassiopeia. He attempted to bargain with the offended god of the sea, Poseidon, for leniency when Cassiopeia slandered the sea nymphs, but he ultimately failed. For enabling his wife's vanity and enraging Poseidon, Cepheus, the King, shared the fate of Cassiopeia, for he also must closely and endlessly circle Polaris. However, the king is not as glamorous or as brilliant as the queen, and his stars can be difficult to locate. Cepheus's five main stars can be formed into an outline of a house, with four of them tracing the walls and one star, named Errai, creating a pitched roof.

## HOW TO FIND IT

**1** The easiest way to discover Cepheus is to use the pointer stars, Merak and Dubhe, in the Big Dipper. Trace that line in the sky from Merak to Dubhe to Polaris and to Caph. About halfway between Caph and Polaris you will find Errai. Errai is just a little brighter than the dimmest star in Cassiopeia's M, named Segin. If you can see all five stars in Cassiopeia, you should be able to find that part of Cepheus.

**2** Arab astronomers did not share the ancient Greek vision that these stars formed a king. They instead imagined the stars in and around Cepheus to be a shepherd and his dog guarding a flock of sheep. The name Errai echoed this other mythology since it means "the shepherd."

**3** Errai is a yellow star about 45 light-years away. This star on the top of Cepheus is not a supergiant star like most of the prominent stars you see in the sky. It is still 40 percent more massive than the Sun, but compared to some of the stars you meet in the winter sky, Errai is downright puny. If it was a little farther away, it would be impossible to find with the naked eye. Your distant, distant descendants will know Errai quite well, however. For about 2,000 years it will be the North Star. Just as Kochab, a star in the Little Dipper's spoon, used to be the North Star thousands of years ago, so Errai will take over that honor from Polaris around A.D. 3200.

## 17 | DELTA CEPHEI
### *A Variable Star*

| CLASSIFICATION: **STAR** | VISIBILITY: **DIFFICULT** |

There are special stars called Cepheid variables (named after the constellation Cepheus) that helped astronomers map the universe. For various reasons the brightness of these stars pulsated regularly, but there was no way to accurately measure their distance from Earth. American astronomer Henrietta Leavitt unlocked their secret when in 1912 she showed that the more luminous the star, the longer it would take to complete one pulsation. Knowing this relationship allowed astronomers to determine how far away Cepheid variable stars were, based on how quickly they dimmed and brightened again. This allowed them to measure the distances to stars within the Milky Way and even those in distant galaxies.

A star in the constellation Cepheus with the name Delta Cephei is the poster child for Cepheid variables. It changes its brightness regularly and goes from dim to bright to dim again about every 5 days. The variability of this star is very slight, and it wasn't even noticed until 1784 when the English amateur astronomer John Goodricke first documented Delta Cephei's strange behavior.

### HOW TO FIND IT

**1** Delta Cephei is one of the dimmest objects described in this book, fluctuating between third and fourth magnitude.

**2** First find the house-shaped pattern of stars that form the constellation Cepheus then find the star on the bottom left corner of the house. Nearby, just outside the house shape, you'll find a fainter star. That is Delta Cephei, an unassuming star that is a whopping 887 light-years away.

**3** Like all Cepheid variables, this star does fluctuate in brightness, but don't expect to find it dim on one night and blazingly bright on the next. Its variability is not noticeable to the naked eye and is barely detectable through professional telescopes. But this inconspicuous fluctuation turned out to be the key in our understanding of the size of the universe.

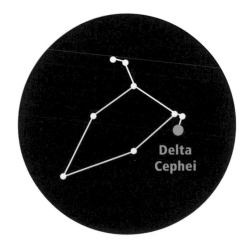

Delta Cephei

**"**For my part, I know nothing with any certainty, but the sight of the stars makes me dream…**"**

Vincent van Gogh, artist

# DRACO

*The Dragon*

CLASSIFICATION: **CONSTELLATION** | VISIBILITY: **MODERATE**

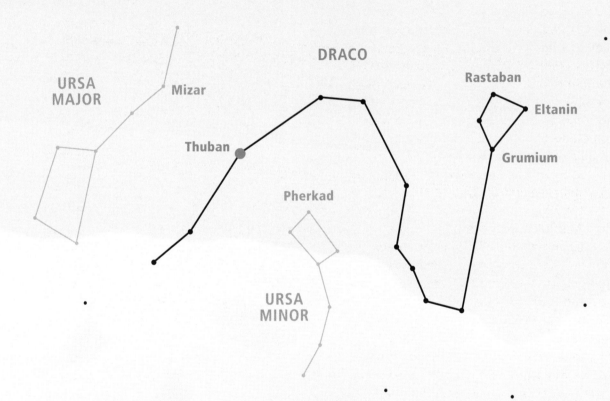

The outline of the long and skinny constellation Draco, the Dragon, resembles a serpent that coils near the constellations Ursa Minor and Cepheus. Draco plays a central role in one of the oldest surviving creation myths, an ancient Babylonian tale called the Enuma Elish. This myth describes an epic battle between good and evil, between the god Marduk and the serpentine monster named Tiamat.

In the beginning there was only darkness. Then creatures emerged and roamed throughout the universe. One of these creatures was a huge she-dragon named Tiamat that was seven miles long from head to tail and ruled over all. She had a mob of other monsters that carried out her bidding and kept everyone in constant terror.

There was a god named Marduk who was said to be able to control the seven winds and feared no monster. Marduk had a giant net, a bow, and arrows, and he made it his quest to defeat Tiamat and free the fledgling world from her tyranny.

When Marduk approached Tiamat, the mighty dragon towered over him. He threw out the net, which was carried by the seven winds to encircle the dragon's body. Tiamat laughed at this and snatched Marduk up in her mighty claw. She then opened up her mile-wide jaws and was about to devour Marduk when he sent the seven winds right down her throat. The winds puffed her up like a balloon while the net held her down. Then Marduk notched an arrow to his bow and let one fly down her open mouth, past her sharp fangs, and into her heart. Tiamat was dead.

Marduk then sliced Tiamat's body into two great parts. One half he threw upward to become the heavens, the other half he laid down as the Earth. Marduk then rounded up the monsters and threw them into the sky to become the stars around the once-mighty Tiamat that you can still see today embodied in the constellation that the Greeks referred to as Draco.

## HOW TO FIND IT

**1** Draco's place in the heavens is moderately easy to locate even though its stars are difficult to outline. First find the Big Dipper and Little Dipper. In between these two more prominent star patterns look for a coiling curve of faint stars that form the dragon's tail. Continue to follow that curve of stars around the tip of the Little Dipper's spoon—that is the dragon's body. The dragon's head is by far the most recognizable feature. Three semi-bright stars make a neat triangle and bear ancient names. They are, in order of brightness, Eltanin, Rastaban, and Grumium.

**2** Even though Draco is a circumpolar constellation (a group of stars that circle the North Star and can be seen all year round), its head stars can get very low in the sky after sunset during the **WINTER** months and become impossible to see. During the **SPRING** evenings Draco's head rises into the northeastern sky. The best time to find the dragon is during the **SUMMER** evenings when Draco soars high above the North Star. During this time of year the three stars in the dragon's head are nearly straight overhead at night. By the **FALL** Draco continues its circle around the Pole and starts getting lower but is still easily visible halfway up in the northwestern sky.

# 19 | THUBAN

## *The Old North Star*

| CLASSIFICATION: **STAR** | VISIBILITY: **DIFFICULT** |

Thuban is blue-white in color and lies about 303 light-years from Earth. Astronomers believe that it gives off about 120 times more light than our Sun. If it was closer, Thuban would definitely be more noticeable and would shine as bright as the brightest stars in the sky. Although dim, Draco's star Thuban has acquired many nicknames, such as "Judge of Heaven," "High Horned One," "Proclaimer of Light," and "Crown of Heaven." Why do astronomers give this insignificant-looking star so much devotion?

Because of precession, Earth's incredibly slow wobble, 5,000 years ago Thuban took its turn as the North Star. Around 2800 B.C. it so perfectly aligned with Earth's North Pole that it did not appear to move at all. It seemed to be an ever-present fixture in the northern sky. The ancient Egyptians definitely noticed. In the construction of one of the great pyramids, they included a long air shaft from the center of the pyramid to the outside. This shaft was designed as a window to allow one view of the sky at one specific angle. Anyone who stood inside the center of the great pyramid could look out and see Thuban through the long window.

## HOW TO FIND IT

**1** Thuban marks a central joint in Draco's long tail, but it is not very bright. In fact, if you have any light pollution in your viewing location, you may struggle to locate it.

**2** That said, the best way to target Thuban is to look halfway between the dimmer star of the tip of the spoon of the Little Dipper (Pherkad) and the middle star on the Big Dipper's handle (and eye-exam star, Mizar). If you see a star in this position, you have found Thuban and have good eyesight and a clear sky.

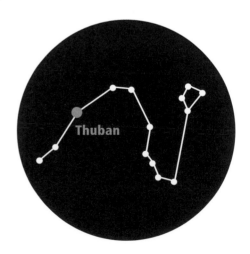

"It's human nature to stretch, to go, to see, to understand. Exploration is not a choice, really; it's an imperative."

Michael Collins, astronaut

# The Winter Sky

While the stars and constellations in the northern sky are visible all year round, the rest of them are best seen during certain seasons. Astronomers call these seasonal constellations. The stars and constellations of winter are some of the brightest and easiest to recognize of the whole year, and they provide some of the best stargazing for the beginner.

To view seasonal constellations, you'll be facing mostly east, south, and west. Since only circumpolar constellations (those visible all year) reside in the northern sky, we won't need to look much in that direction. Over the course of a night seasonal stars will rise in the east, travel up and to the right until they reach their highest point above the horizon in the southern sky. Then they will start heading down and to the right and toward the western horizon until they set.

Once the New Year dawns, the constellations of winter take center stage and cover the eastern and southern quadrants of the sky. To the ancient Greeks this region was filled with mythological creatures interwoven into one large legend. There is one particular entire scene complete with love, love scorned, and death. Would you believe that, in those stars, there is a giant hunter being trampled by a charging bull that has seven women on its back, while nearby two hunting dogs are chasing after a unicorn and a bunny rabbit down by the river?

Learning to find all of these constellations might sound difficult, but we'll go step by step, myth by myth, and explore the placements of the stars and constellations of the winter sky. Along the way we will meet Orion, the Hunter; Taurus, the Bull; Canis Major, the Big Dog; Canis Minor, the Little Dog; Gemini, the Twins; and Auriga, the Charioteer. We'll also take closer looks at the super stars that dwell in these winter constellations. By the end of winter you'll be able to identify them all!

# Winter Sky Constellations

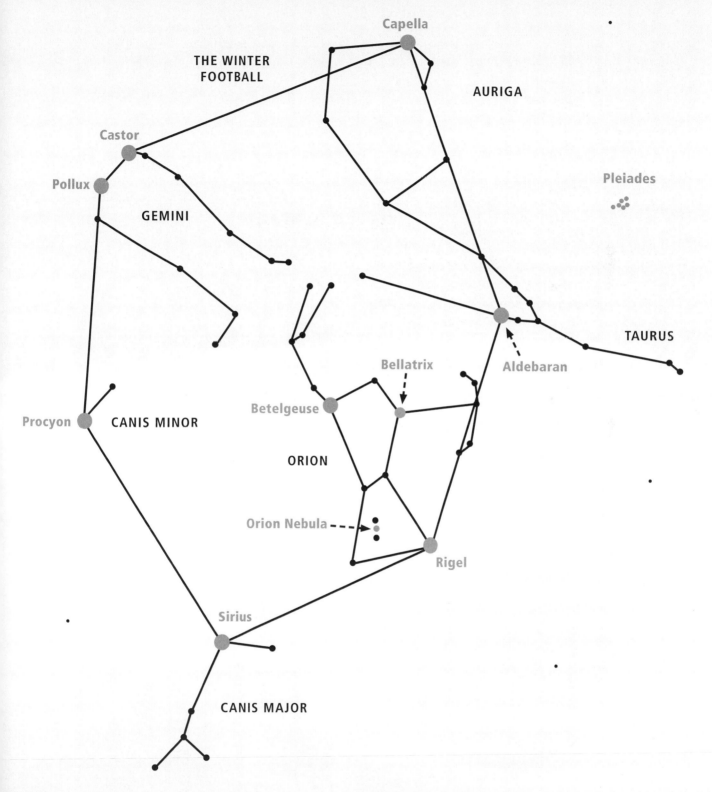

# 20

THE WINTER FOOTBALL

CLASSIFICATION: **ASTERISM**   VISIBILITY: **EASY**

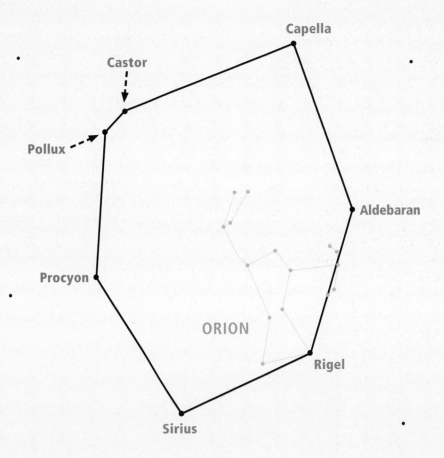

If you can brave the cold, the winter night sky holds the most stellar gems of any season. Eight of the twenty brightest stars in the entire sky, a ring of sparkling multicolored jewels encircling the constellation Orion the Hunter, shine every winter. These brightest winter stars are often called the Winter Circle or Winter Hexagon. You can always see these dazzling stars gracing the skies during the playoffs of the National Football League (NFL) and around the time of the Super Bowl. If you connect the dots on these stars in the right way, maybe you can imagine a giant football—what I like to call the Winter Football—flying through the air.

At 65 degrees long and 40 degrees wide the Winter Football covers almost half of the entire southern sky. Pointy on two ends, rounded in the two middle halves, this Winter Football is the huge star pattern of the season.

The previous star chart shows what the sky looks like on a normal winter evening in January and February around 8 or 9 p.m. The Winter Football is outlined and labeled. You can use this asterism as a guide to find many of the stars and constellations inside. To the ancient Greeks this region was filled with mythological creatures interwoven into one large legend. Never fear. I will walk you through the stories told in the stars of the winter sky.

## HOW TO FIND IT

**1** To trace out the Winter Football, start by locating Sirius, the Dog Star, the brightest star in the night sky. It can be found at the pointy end of the football that is nearest to the ground.

**2** Next, going clockwise around the football, you will find Procyon, the Little Dog Star, followed by the Gemini Twins' head stars, Pollux and Castor.

**3** From there, go to bright Capella, which marks the upper point of the football and is often very high in the sky.

**4** Make a quick turn down toward the horizon and you'll trace the other side of the football toward the Bull's Eye star, Aldebaran.

**5** Continue around to Rigel, Orion's left foot, and then head back to Sirius. By doing this, you'll have traced the entire Winter Football!

# 21

# ORION

*The Hunter*

| CLASSIFICATION: **CONSTELLATION** | VISIBILITY: **EASY** |

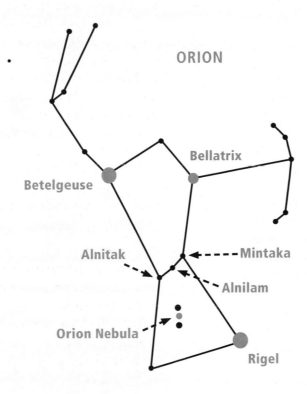

ORION

Betelgeuse

Bellatrix

Alnitak

Mintaka

Alnilam

Orion Nebula

Rigel

Orion is the constellation that conjures the deepest imagination and wonder with just one glance. Something about the placement of the stars ties the entire picture together. Almost every culture in the ancient world associated these stars with a hunter, a giant, or an all-around he-man. His origin story in the Greek myths is murky at best and no storyteller can seem to agree as to where he came from or how he got to be so tough. Orion typically appears in tales that require a hunter, and in these stories he proves himself to be the best around.

Orion is very easy to see in the night sky and, unlike many constellations, the hunter is easy to imagine as a hunter. It's best to picture him facing you with his right arm raised high holding a sword and his left arm bent and holding a shield. But Orion's Belt—made up of the stars named (from left to right) Alnitak ("the girdle"), Alnilam ("the string of pearls"), and Mintaka ("the belt")—ties the whole constellation together. Orion's Belt is a great landmark in the heavens; not only does it identify the giant hunter constellation but it helps point us to other stars and constellations in the winter sky.

Below the belt are two stars that mark Orion's feet (bright blue Rigel and fainter Saiph), and above the belt are his shoulder stars (Bellatrix and Betelgeuse). At the top of the constellation is a very faint star marking his head, named Meissa. In all of the legends about Orion he is portrayed as being extremely brave, strong, fearless...but not very bright. And the dim star marking his head displays that appropriately.

## HOW TO FIND IT

**1** To find Orion simply look for three stars of average brightness that form a tight line in the WINTER skies. Locate them and you have found Orion's Belt. There is no other star pattern visible to the naked eye quite like it.

**2** You can easily identify Orion every evening between JANUARY and APRIL. When he rises in the east-southeast in JANUARY, he will seem a little unbalanced. He looks like he's tipped over on his side and the belt stars will be like a vertical line pointing straight down to the horizon. As he reaches his highest point in the southern sky after dark in FEBRUARY, Orion rights himself. Standing about halfway up in the southern sky, the outline of Orion's stars are so dramatic that he looks to be perched over the Earth and commanding the heavens.

**3** As we get to MARCH, when Orion sets in the west-southwest, he once again seems to tip over just before he sets below the horizon. Once APRIL turns to MAY, Orion is no longer visible in the evening sky. But by FALL you can see his trademark belt of three stars in the morning sky, rising in the east just before sunrise.

# 22 | BETELGEUSE

## *The Armpit*

| CLASSIFICATION: **STAR** | VISIBILITY: **EASY** |
| --- | --- |

The most infamous star name in all astronomy is Betelgeuse (because it is most commonly pronounced *Beetle-juice*). Betelgeuse is actually a shortened and edited version of what Arab astronomers called this star: Ibt al-Jauza. Astronomy historians believed that the name originally meant "Armpit of the Central One." After looking at the common illustration of Orion, you would think Betelgeuse would correspond simply with Orion's right shoulder. However, in most depictions of Orion he is raising a club high in the air and thus exposing his armpit.

Betelgeuse is much different than Orion's other bright stars. Betelgeuse is a red supergiant star, and to the naked eye it definitely appears more orange than other stars. When you compare Betelgeuse to Orion's other stars you will see the difference. Bellatrix, Rigel, and the three belt stars are all blue or white. When you look back at Betelgeuse after peering at Orion's other bright stars, you'll agree it is much redder than the others.

Betelgeuse is about 640 light-years from Earth and it is also humongous. If it were our Sun, its volume would stretch beyond the orbit of Jupiter. That means the Earth would orbit inside it! Betelgeuse is so big that it is one of a few stars on which astronomers can map features.

But maybe the coolest thing about Betelgeuse is that it will explode. It's so massive that when it dies, it will create a bright supernova. In fact, this explosion should be so bright that you will be able to see it in the daytime. Astronomers do not know when it will go supernova. It could be tomorrow or it could be centuries from now. But keep checking on Betelgeuse, because maybe one night you'll witness Orion's armpit explode!

### HOW TO FIND IT

1. After you locate the constellation Orion, look for two bright stars above Orion's Belt that could stand in as his shoulders.

2. Since Orion is facing us, look at his right shoulder (on your left) and it should be much redder than his left shoulder (on your right). That is Betelgeuse, and it may seem embarrassing at first, but that is actually Orion's armpit. He doesn't mind you staring!

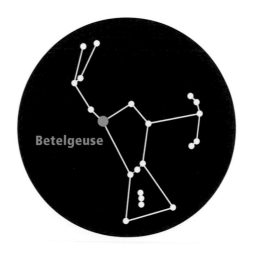

# 23 | BELLATRIX

## *The Beautiful*

| CLASSIFICATION: **STAR** | VISIBILITY: **EASY** |

Orion's left shoulder, a bright blue-purple star, is named Bellatrix. Bellatrix, which is about 250 light-years from Earth, has one of the deepest blue colors of any bright star. This color tells astronomers that it is much hotter than Betelgeuse. Bellatrix has a surface temperature of about 40,000 degrees Fahrenheit while Betelgeuse is a relatively cool 5,400 degrees.

The origin of the name Bellatrix comes from ancient Greek and Roman, and it means either "beautiful warrior woman" or "Amazon star." Legend had it that this star rendered all women born beneath it lucky and talkative. More specifically, this star was linked to great Amazon women who were strong, well-spoken, assertive, and tough. Bellatrix has an Arabic name as well. Because it is the first star in Orion to rise above the horizon, it is called Al Murzim, "the herald of Orion."

Native groups living in the Amazon River basin in South America gave human characteristics to individual stars. They didn't make entire constellations, but they did dwell on the comings and goings of Betelgeuse and Bellatrix. To them each star was a person sitting in a canoe. Bellatrix was a young boy swiftly paddling through the waves with ease. Betelgeuse was an old man struggling with all of his might to keep up.

## HOW TO FIND IT

**1** You can find Bellatrix shining down on you from Orion's left shoulder (on your right).

**2** First find Orion's Belt, then look at the two stars marking his shoulders. Betelgeuse is the redder star on your left and Bellatrix is the bluer star on your right.

# 24 | RIGEL

## *The Left Foot*

| CLASSIFICATION: **STAR** | VISIBILITY: **EASY** |
|---|---|

The brightest star in Orion, and the seventh brightest star in the entire sky, is named Rigel. This name can be loosely translated as "left foot of the central one." It twinkles blue-white in color and makes a great contrast to orange Betelgeuse on the other side of the constellation.

Rigel figures quite prominently in the ancient Greek legend surrounding Orion's untimely death. Orion bragged that he was such a formidable hunter that he could wipe out all life on Earth. To humble the mighty hunter, the gods thought it would be ironic if Orion was killed by a tiny, almost insignificant creature. He was stung on his left heel by a scorpion—now embodied in the summer constellation Scorpius—and the star Rigel was said to mark the site of the fatal sting.

Rigel is about 860 light-years away but is still one of the brightest stars in the night sky. That means it must be incredibly huge and immensely luminous. Not only does Rigel have a diameter 100 times greater than the Sun but it is also a hot star with a surface temperature of over 21,000 degrees Fahrenheit. This combination of size and brilliance means that Rigel may put out as much as 279,000 times more light than our Sun produces.

Astronomers believe there could be three to five stars in the Rigel system, which means that the star you see with the naked eye is merely the brightest component of a system of suns circling a common center. Rigel is so radiant that it is difficult to see any other stars that may be orbiting it, but astronomers have found direct evidence of at least two other stars near Rigel, and there may be more yet to discover.

### HOW TO FIND IT

**1** Look for blue-white Rigel, the sting on Orion's left foot.

**2** From Orion's Belt you will see two stars that could form his feet dangling below. Rigel is the brighter foot star and can be found on his left side (your right).

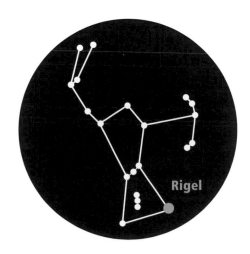

# 25 | THE ORION NEBULA

## *The Great Star Factory*

| CLASSIFICATION: **NEBULA** | VISIBILITY: **MODERATE** |

Hanging from the center star of Orion's Belt are three "stars" that form a small sword. The middle of these three appears to be fuzzy in a dark sky. This is not a star, but a star cloud called the Orion Nebula, or M42, a cloud of gas and dust 24 light-years wide that is creating new stars. In this case, this nebula is a huge star factory.

Even though it is about 1,344 light-years away, the Orion Nebula—which has enough material in it to create about 2,000 Suns—is the closest large star-forming region to Earth. Sometimes this nebula can appear fuzzy because you can actually see its nebulosity, the light from the clouds of gas and dust traveling 1,344 light-years to you.

### HOW TO FIND IT

**1** To find the Orion Nebula simply find the three stars that make Orion's Belt. When you look below the belt, you may discover another one, two, or three, fainter stars in a line. These three stars make up Orion's sword that is hanging from his belt. The middle star of the three is the Orion Nebula.

**2** The Orion Nebula is the second brightest object below the belt. If you can only see two stars below Orion's Belt, the brighter one at the bottom is a star called Nair al Saif. The Orion Nebula is just half a degree above it.

**3** You can see a little more detail in the nebula by using a pair of binoculars. And through a backyard telescope you can resolve individual stars in the Orion Nebula that are surrounded by a gray cloud of material.

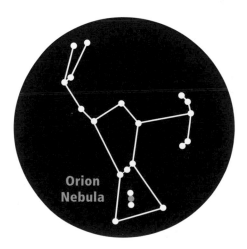

Orion
Nebula

# 26

TAURUS

*The Bull*

CLASSIFICATION: **CONSTELLATION** | VISIBILITY: **EASY**

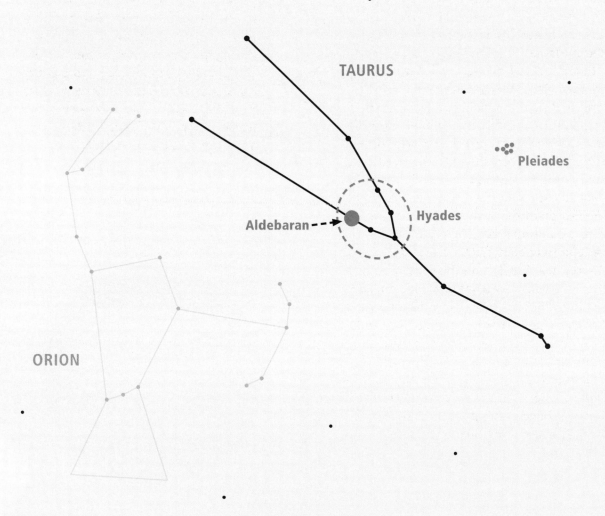

TAURUS

Pleiades

Aldebaran

Hyades

ORION

According to Greek mythology, Taurus, the Bull, was turned into a constellation by the gods in order to protect the Seven Sisters (the Pleiades) from the unwanted advances of Orion. The Seven Sisters got to sit pretty on Taurus's back while Orion fended off the charging beast.

Taurus is one of the most recognizable zodiac constellations in the sky and is arguably the oldest constellation invented. Most astronomers and historians agree that a drawing of the constellation of the Bull is depicted in one of the oldest works of human art found, deep down in a cave near Lascaux, France. This ancient painting looks suspiciously like the stars in the constellation Taurus, the Bull. Taurus has long horns pointing to the left. The spots in the face correspond to the Hyades star cluster (in the face of Taurus—easily seen with the naked eye). And the seven dots in a tight clump represent the Seven Sisters star cluster. This cave painting is actually a star chart—and it was painted about 17,000 years ago!

Taurus, the Bull, lies on the imaginary path in the sky called the zodiac on which the Sun, the Moon, and all the planets appear to move. The zodiac was the earliest form of a calendar, marking the movement of the Sun throughout the year. In ancient days, Taurus held the most important spot—it signaled the return of spring. As a signal of the seasons, the stars of Taurus may have inspired the ancient artist to paint the image of the bull in Lascaux.

## HOW TO FIND IT

**1** Look for a V shape of five stars to the right of Orion. The V marks Taurus's head. The fainter stars extending out from the V form two long horns.

**2** If you are still having trouble finding Taurus, use the three stars that mark Orion's Belt as pointer stars. Connect the three dots on Orion's Belt and continue this line of sight to the right about 20 degrees (two widths of your fist at arm's length). This will take you to a spot just under a bright, orange-colored star. That star is Aldebaran, or the Bull's Eye, which marks the left side of the V shape of stars that is the bull's face.

# ALDEBARAN

*The Bull's Eye*

| CLASSIFICATION: **STAR** | VISIBILITY: **EASY** |

Aldebaran is the brightest star in the constellation Taurus, the Bull. It is super easy to identify as it stands out as the Bull's Eye, the brightest—and reddest—star in the Bull's V-shaped face. Aldebaran is more than 500 times more luminous than our Sun. If Aldebaran was our Sun, its outer reaches would stretch all the way to the orbit of Mercury.

Although Aldebaran looks to be a part of the larger cluster of stars in the bull's face, the Hyades, it is not. At only 65 light-years from Earth, Aldebaran is more than twice as close as that cluster. It just happens to be in the same line of sight as those more distant stars.

Despite Aldebaran's sinister, red appearance, this star was a sign of good fortune. More than 5,000 years ago, the ancient Persians designated Aldebaran as one of the Four Royal Stars, the Guardians of the Sky. Each Royal Star signaled a season of the year, and Aldebaran's proximity to the Sun during the springtime made it the herald of spring. Ancient Hebrews revered Aldebaran as God's Eye. They also called it Aleph or even just A, referring to the first letter of the Hebrew alphabet. The name Aldebaran is an Arabic word loosely translated as "The Follower." What is Aldebaran following? Just continue that line from Orion's Belt past Aldebaran and you will find the most impressive cluster of stars, the Pleiades, or Seven Sisters.

As the night goes on, these stars will seem to move from left to right across the sky. This movement gave ancient stargazers the illusion that Aldebaran was following the Seven Sisters.

### HOW TO FIND IT

**1** To find this star, connect the stars of Orion's Belt and keep going to the right.

**2** After traveling about 20 degrees, this line will take you just below Aldebaran, which glows with an orange light.

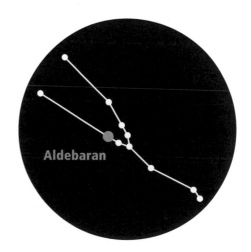

Aldebaran

# 28 | THE HYADES STAR CLUSTER

| CLASSIFICATION: **STAR CLUSTER** | VISIBILITY: **MODERATE** |
| --- | --- |

If you look carefully at the V-shaped face of Taurus, the Bull, you may notice that there are several fainter stars scattered around. Most of these stars that you can see with the naked eye are part of an open cluster called the Hyades.

Open clusters are dozens, hundreds, or even thousands of stars that were all formed from one massive cosmic cloud. All the stars within a cluster are approximately the same age and distance from Earth. Astronomers have found more than 1,100 open clusters in the Milky Way alone, and they have observed other star clusters in nearby galaxies.

The Hyades is the closest open star cluster to Earth, and it is made up of about 250 stars that all lie about 150 light-years away. Five to ten of the stars can be seen with the naked eye since they are either third or fourth magnitude in brightness. These stars all likely formed from the same nebula, which is a giant cloud of gas and dust.

According to Greek mythology, the Hyades were five daughters of the god Atlas (who is most famously depicted as holding up the world) and half sisters to the Pleiades. Their appearance in the sky seemed to correspond to the rainy season, and so they were often pictured to be crying their eyes out while mourning the death of their brother, Hyas. Their tears became the spring rains.

## HOW TO FIND IT

1. To find the Hyades, use Orion's Belt as a guide. Connect the three stars in Orion's Belt and continue this line to the right about 22 degrees to the heart of the Hyades, which, along with Aldebaran, fill in Taurus's face.

2. If you extend your right arm and spread your thumb and fingers, then place your thumb on Orion's Belt, your pinkie should reach to the Hyades.

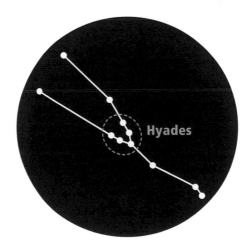

# THE PLEIADES

## *or Seven Sisters*

| CLASSIFICATION: **STAR CLUSTER** | VISIBILITY: **EASY** |

The Pleiades, also called the Seven Sisters, is the most famous and impressive naked-eye star cluster in the sky. At first glance the stars in the Pleiades look more like a little cloud, but upon closer examination you may detect five, six, or seven individual stars. However, this cluster is actually a large group of stars moving together in space, about 400 light-years from Earth. Astronomers believe them to be very young and hot stars—formed only 500 million years ago (compared to our Sun, which is 5 billion years old). With binoculars you can see approximately 50 stars in this cluster, and if you carefully scan the boundaries of the Pleiades with a telescope you can count all 500 of them.

People with good eyesight think the Seven Sisters resemble a very small dipper, but the ancient Greeks thought the entire group of stars resembled the outline of a dove in the sky. Legend has it that the sisters were trying to flee the amorous advances of Orion and only escaped by divine intervention. The gods took pity on the sisters and turned them into doves. They flew away from Orion and landed up in the sky where they sit today.

That said, the Pleiades are not a constellation of their own. They are part of the larger constellation Taurus, the Bull. When Orion obtained his place in the sky, he continued to harass the Seven Sisters. So the gods established Taurus to reside between the two, forever protecting the Sisters from Orion.

## HOW TO FIND IT

**1** To locate the Seven Sisters, use Orion's Belt as a pointer. Draw a line through the three stars in the belt and continue that line to the right. This will take you to the face of Taurus, the Bull, but be sure to keep going.

**2** A little more than 10 degrees past Taurus you will run into the Pleiades. It's quite a large hop from Orion's Belt to the Seven Sisters—roughly 35 degrees in the sky—but once you identify them, you'll know why the ancient world was so enamored with these stars.

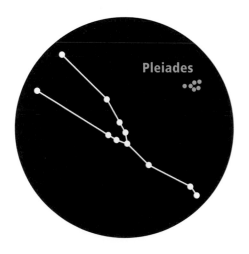

"In the final analysis, our most basic common link is that we all inhabit this small planet. We all breathe the same air. We all cherish our children's future. And we are all mortal."

John F. Kennedy, American president

# 30

# CANIS MAJOR

*The Big Dog*

CLASSIFICATION: **CONSTELLATION** | VISIBILITY: **EASY**

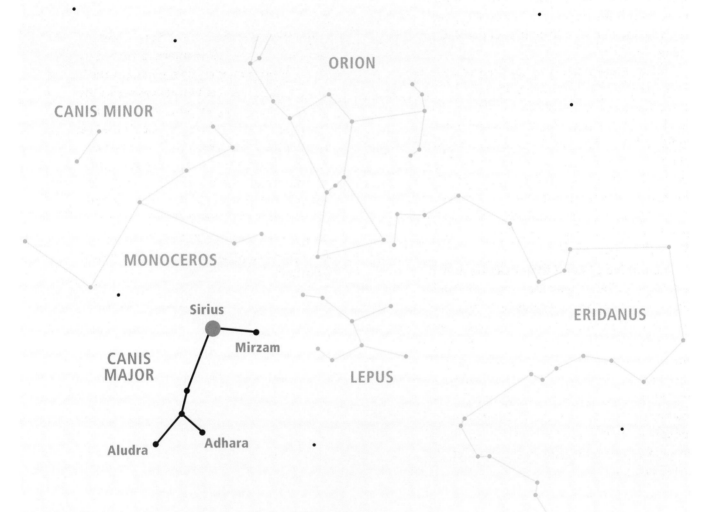

ORION

CANIS MINOR

MONOCEROS

Sirius

Mirzam

CANIS
MAJOR

LEPUS

ERIDANUS

Aludra     Adhara

The ancient constellation Canis Major ties in with the Greek mythology of Orion and his placement in the sky. According to the legend, after Orion was stung by a scorpion (the summer constellation, Scorpius) and died, he asked the gods if he could bring his two favorite hunting dogs with him into the sky to help fend off the menacing bull, Taurus. The gods agreed to this, and you can find the big dog, Canis Major, and the little dog, Canis Minor, to the left of Orion.

Unfortunately, Canis Major doesn't seem to be helping his master much since he is being distracted by other constellations in the winter sky. He is busy chasing after a unicorn (the constellation Monoceros) and a bunny rabbit (the constellation Lepus) by the banks of the cosmic river (the constellation Eridanus). The unicorn is between Canis Major and Canis Minor. Meanwhile, the rabbit hops below Orion's feet, and the river starts flowing down by Orion's foot star, Rigel.

The outline of Canis Major can be drawn several ways, but it may look more like a beagle or a Scottie than a vicious hunting dog. The astounding star Sirius can act as an eye with a fainter nose-star called Mirzam sticking out to the right. Below Sirius are the dog's neck and two stubby legs (the stars Adhara and Aludra). Canis Major's body gets fainter toward his rear end on the left side of this constellation. But you may be able to detect his back legs in a neighboring constellation called Puppis.

## HOW TO FIND IT

**1** Use Orion's Belt as a guide to find Canis Major. Connect the dots on the three belt stars from right to left. Then extend that line to the left and travel about 20 degrees in the sky. That will take you to the dog's nose and the brightest star in the sky, Sirius.

**2** From the mainland United States and any country in mid-northern latitudes, Canis Major never rises very high in the sky. In **JANUARY** the big dog rises in the southeast after sunset and, as the night rolls on, seems to scamper just above the southern horizon. In **MARCH** look for Canis Major about 30 degrees up in the southern sky. And by **APRIL** it's only visible for a short time in the southwestern sky before it sets. Knowing its lower position in the sky, you can also find Canis Major at the bottom point of the Winter Football.

**31** | # SIRIUS

## *The Dog Star*

| CLASSIFICATION: **STAR** | VISIBILITY: **EASY** |

The brightest star visible from Earth (other than the Sun) is Sirius. The ancients called it the "scorcher," and it lives up to its name by blazing almost twice as bright as the second brightest star, Canopus, which is found in the southern constellation Carina. (Carina is not visible from the mainland United States.) Sirius is so dazzlingly bright because it is relatively close to Earth. It is only 8.6 light-years away, making it the seventh closest star beyond the Sun.

Sirius has fascinated people around the globe, and it plays a prominent role in countless cultural myths. Ancient Egyptians worshipped it as the King of Suns and based their calendars on its movements. The rising and setting of Sirius told the Egyptians when to plant, when to harvest, and when the Nile typically flooded. In Hindu mythology Sirius was a hunter and the father of 27 daughters, represented in the 27 daily phases of the Moon that were visible throughout its cycle.

As the brightest star in the constellation Canis Major, Sirius has acquired the nickname, "the Dog Star." In fact, the phrase "dog days of summer" originates in ancient Greece. During the hottest part of summer the Sun stood near Sirius in the sky. And during the hottest months the Greeks claimed that the unseen presence of Sirius added to the intensity of the Sun's heat and caused the dog days of summer.

## HOW TO FIND IT

**1** Finding Sirius in the sky is about as easy as it gets. Just look for the brightest star in the night sky. Only the planets Venus, Jupiter, and Mars can shine brighter, but there is an easy way to tell them apart from Sirius. Stars twinkle much more than planets, and Sirius often twinkles red, white, and blue when it is low in the sky. If you see a suspiciously bright star noticeably twinkling in the winter sky, then you've found the Dog Star, Sirius.

**2** You can also use the stars of Orion's Belt for guidance. Just connect the three stars in Orion's Belt and continue that line of sight down and to the left. After traveling about 20 degrees in that direction, you will run into Sirius.

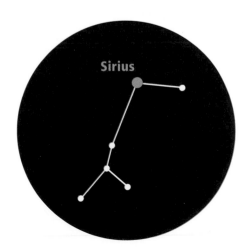

# "Magnificent desolation."

Buzz Aldrin, astronaut, upon setting foot on the Moon

# CANIS MINOR

## The Little Dog

| CLASSIFICATION: **CONSTELLATION** | VISIBILITY: **EASY** |

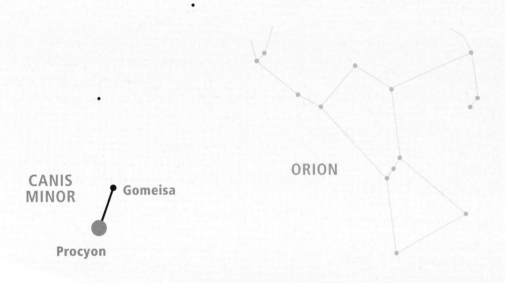

ORION

CANIS
MINOR

Gomeisa

Procyon

Sirius

CANIS
MAJOR

Canis Minor is Orion's smaller hunting dog, and it is probably one of the most underwhelming constellations in the entire night sky. With the naked eye you can often only make out two stars in the entire star pattern of Canis Minor: Procyon and Gomeisa.

The gods felt a little sorry about Orion's predicament in the sky. There he was with his belt of three stars just trying to find love with the Seven Sisters. He had fruitlessly chased them on Earth and now continued that pursuit in the heavens. On top of that there was the Bull standing in his way, threatening to trample him every night. He needed help, and so he appealed to the mercy of the gods and was magically granted two hunting dogs to help him with the Bull. Canis Major, the bigger and more helpful of his canine companions, seems up to the task; while Canis Minor seems to be less than an effective assistant. He is tiny compared to the Bull, and with Orion bearing the brunt of the attack, is too far from the action to be of much help. It almost looks like Orion is defending Canis Minor.

## HOW TO FIND IT

**1** To find Canis Minor, first find the three stars in Orion's Belt then continue that line of sight down and to the left for 20 degrees until you reach Sirius, the Dog Star, in the constellation Canis Major.

**2** Then make a 90-degree turn to the right from Sirius and travel another 25 degrees. There you will discover another very bright star called Procyon. This is also called the Little Dog Star and it marks the center of the body of Canis Minor.

**3** Just 4 degrees to the right is a fainter, second magnitude star called Gomeisa that stands in for the little dog's head. Congratulations, you have now observed the entire constellation of Canis Minor!

## 33 | PROCYON

### *The Little Dog Star*

| CLASSIFICATION: **STAR** | VISIBILITY: **EASY** |

Procyon is the eighth brightest star in the sky and shines with a sparkling white light. It is similar in size to Sirius and is also relatively close, residing about 11.5 light-years from Earth.

Other cultures created mythologies around just this star. Babylonians called it Nangar, the carpenter who helped construct the heavens above. Hawaiians used Procyon as a navigational aid for traversing the Pacific on their epic sea voyages. And in Inuit cultures it was called Sikuliarsiujuittuq, the hunter who, because of his tremendous girth, should not go onto newly formed ice.

Our name, Procyon, comes from the Greek. It means "before the dog," since it rises above the eastern horizon just before Sirius, the Dog Star, comes up in the southeast. Because of its placement in the sky, Procyon goes higher in the south, stays up longer in the sky than Sirius, and sets after the Dog Star. Both dog stars, Sirius and Procyon, have a hidden secret. Each of these stars may look like a simple, bright white star, but each has a small companion, called a white dwarf star, revolving around it. These white dwarf stars have nearly the mass of our Sun but are only about 1 percent of its size. That means these companion stars are ultra-dense. These dwarf stars are so small in the sky you need a large telescope to make them out.

### HOW TO FIND IT

**1** To find Procyon, you just follow the same directions that brought you to Canis Minor. Travel down and to the left from Orion's Belt for 20 degrees until you come to the brightest star in the sky, Sirius.

**2** Then take a 90 degree right turn and go 25 more degrees until you reach another impressively bright star. That's Procyon, the body of the little dog in Canis Minor.

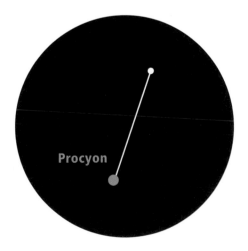

"We cannot predict the new forces, powers, and discoveries that will be disclosed to us when we reach the other planets and set up new laboratories in space. They are as much beyond our vision today as fire or electricity would be beyond the imagination of a fish."

Arthur C. Clarke, science fiction author

# GEMINI

## *The Twins*

| CLASSIFICATION: **CONSTELLATION** | VISIBILITY: **EASY** |

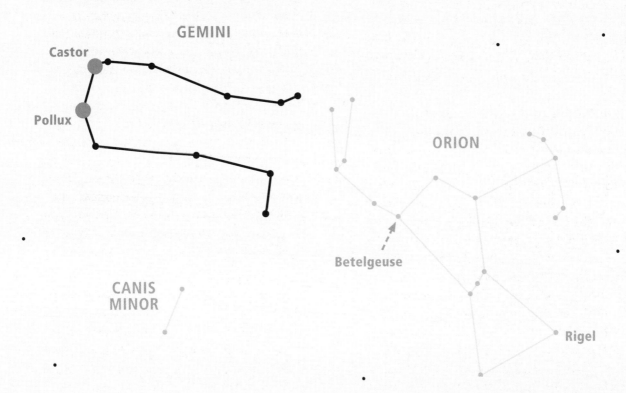

GEMINI

Castor

Pollux

ORION

Betelgeuse

CANIS MINOR

Rigel

The constellation of Gemini, the Twins, is one of the oldest and easiest to recognize star patterns in the sky. Two bright stars named Pollux and Castor stand in for the heads of the twins. However, once you find them and you look a little closer, you will notice that they are not identical twin stars. Pollux is yellow-orange in color and slightly brighter, whereas Castor is blue-white and noticeably dimmer. Pollux is a giant orange star that is about 9 times wider and 43 times more luminous than the Sun. It is situated about 34 light-years away from Earth. Castor, which lies about 51 light-years from Earth, is not just a solitary blue-white star but is in fact a system of six stars that revolve around each other.

The word *Gemini* comes from the Latin meaning "twins." But it was the ancient Greeks who spun the most fanciful tale about these brothers. According to ancient Greek mythology Pollux and Castor shared the same mother, but one was mortal (Castor) and the other was the son of Zeus (Pollux). They grew up developing the greatest bond of friendship. One evening the brothers attended the double wedding of their male cousins (also twins) who were marrying, you guessed it, twin girls! Before the ceremony began, Pollux and Castor accidentally went into the wrong tent where the twin girls were readying themselves. Well, their eyes met and the twins fell helplessly in love with the twin brides-to-be. The foursome were about to make a quick and romantic getaway from the wedding when the two grooms discovered their plans and stopped them. A terrible fight ensued in which Castor was killed. In Pollux's rage, he killed the two cousins in revenge.

After the brawl Pollux was so saddened that he wished he were dead. He pleaded with the gods to kill him so that he could be with his brother forever in the afterlife. The gods were so moved by Pollux's feelings that they granted his request and immortalized the twins together in the sky to be a sign of fraternal love.

## HOW TO FIND IT

 To find Gemini, look for their head stars, Pollux and Castor, using Orion's two brightest stars, Rigel and Betelgeuse, to point the way. Start at Rigel and then draw a line toward Betelgeuse. Keep going and continue that line for about 30 degrees until you find two bright stars, Pollux and Castor. Once you find the head stars of the twins, follow two lines of fainter stars for about 20 degrees (or 2 times the width of your fist at arm's length) that point toward the constellation Orion. With a little imagination you may be able to picture the twins dancing atop Orion's club.

2 Gemini is best found on **WINTER** evenings, and it rises in the east-northeastern sky starting in **DECEMBER**. By **JANUARY** and **FEBRUARY** this constellation will quickly ascend to lofty heights and ride very high in the southeast. During **MARCH** and **APRIL** you can catch Gemini halfway up in the western sky standing above Orion before he sets. At the end of **MAY**, this constellation says goodbye as it gets lost in the glare of sunset.

# 35 | POLLUX AND CASTOR

## Double Trouble

| CLASSIFICATION: **STARS** | VISIBILITY: **EASY** |

Pollux and Castor are the brightest stars in the constellation Gemini, the Twins. However, once you find them and you look a little closer, you will notice that they are not identical twin stars. One star is yellow-orange in color and slightly brighter—that's Pollux—and the other is blue-white and noticeably dimmer—that's Castor.

Pollux is a giant orange star that is about 34 light-years away. It is about 9 times wider and 43 times more luminous than the Sun. Castor lies about 51 light-years from Earth but is more than meets the eye. It is not just a solitary blue-white star but is in fact a system of six stars that revolve around each other. That means if you lived on a planet orbiting Castor, you'd have six suns in your sky!

In African mythology Pollux and Castor were known as the Wise and Foolish Antelope. Other cultures called them the Two Peacocks, the Two Kids, or the Giant's Eyes. An old Australian myth ties in the two bright stars of Gemini with the star Capella in the constellation Auriga. In that story, Castor and Pollux are two huntsmen named Yurree and Wanjil. They are hunting the elusive kangaroo named Purra (represented by Capella). During the summer when these stars are below the horizon, it is said that the two hunters finally catch the kangaroo and kill it. They then cook his meat over a fire and cause waves of heat to rise above the ground like a shimmering haze.

## HOW TO FIND IT

1. You can find Pollux and Castor easily by using Orion's two brightest stars, Rigel and Betelgeuse, to point the way. Start at Rigel and then draw a line toward Betelgeuse.

2. Keep going past Betelgeuse and continue that line farther for another 30 degrees and you will quickly be able to identify the only two similarly brilliant stars in that part of the sky: Pollux and Castor.

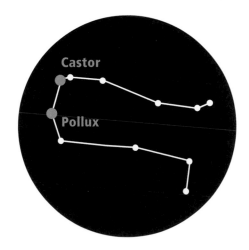

"Then in the middle of all stands the Sun. For who, in our most beautiful temple, could set this light in another or better place, than that from which it can at once illuminate the whole?"

Nicolaus Copernicus, mathematician and astronomer

# AURIGA

## *The Charioteer*

| CLASSIFICATION: **CONSTELLATION** | VISIBILITY: **MODERATE** |

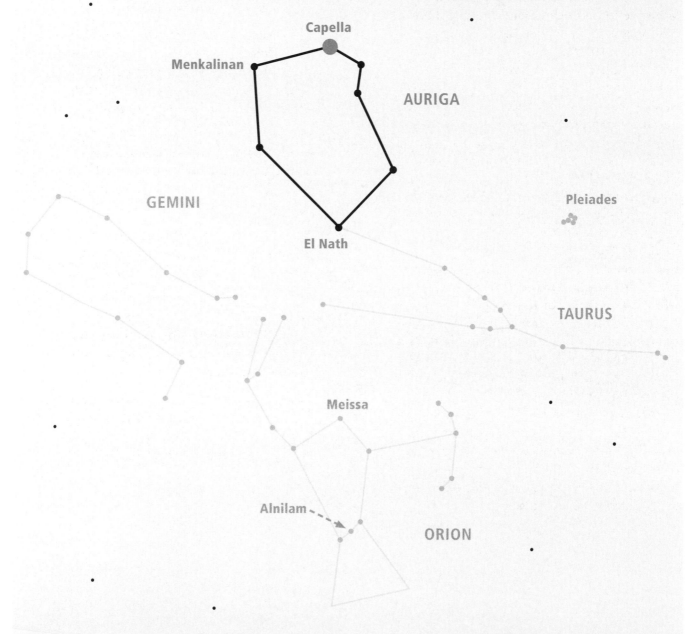

The large winter constellation Auriga features one of the brightest stars in the sky (Capella) and several others that are visible to the naked eye. It was long associated with chariots and charioteers in both Greek and Chinese mythology.

One Greek legend refers to the constellation Auriga as Erichthonius, the son of the gods Vulcan and Minerva. Erichthonius was born deformed and could not walk well. To remedy his situation, he invented the four-horse chariot to get him around the kingdom. He was so respected for his invention that he became the fourth king of Athens. Erichthonius also had a soft spot for crippled or injured animals, his favorite being a little she-goat that is represented by the star Capella. In the sky we are supposed to see Erichthonius holding little Capella and two other goats as they race around the heavens.

Unlike Orion, who is pictured facing us stargazers, it's best to picture Auriga facing away with his back to us. A star called El Nath, which is also the tip of Taurus's left horn, is Auriga's left heel. The second brightest star in this constellation marks his right shoulder. This star is called Menkalinan, a name that comes from the Arabic words meaning "shoulder of the rein-holder."

Chinese astronomers saw a similar thing as the Greeks. In Chinese mythology these stars in Auriga were part of the five chariots of the celestial emperors. In other versions many of the stars of Auriga were chairs of the emperors or posts to tie up the horses.

## HOW TO FIND IT

1. It's not easy to see a guy holding three goats in this star pattern. The constellation itself looks more like a squished pentagon. The surest way to know where Auriga stands is to use the nearby stars in the constellation Orion as guides.

2. Draw a line upward between Orion's middle belt star, Alnilam, and his dimmer head star, Meissa. If you keep going, after about 30 degrees this line of sight will direct you to the center of Auriga.

3. Auriga is also near Taurus, the Bull. These two constellations actually share a star named El Nath, which both marks the tip of the left horn of the Bull and Auriga's left heel.

4. Auriga rises in the northeast earlier than the rest of the WINTER constellations, and it starts becoming visible on OCTOBER and NOVEMBER evenings.

5. By DECEMBER, JANUARY, and FEBRUARY it rides very high in the sky and can even perch itself at the zenith point, straight overhead.

6. On MARCH and APRIL evenings, Auriga is about halfway up in the western sky. It can linger on into MAY and shine just after sunset in the northwest.

## 37 | CAPELLA

*A Notorious Twinkler*

| CLASSIFICATION: **STAR** | VISIBILITY: **EASY** |

Capella is one of the brightest stars in the sky and is similar to our sun in temperature, which means it should look about the same color: yellowish-white. Arab astronomers called this star the Driver, the Singer, and the Guardian of the Pleiades. It was called the Heart of Brahma in India. And in South America this star was one of the favorites among shepherds, who called it Colca. Capella is by far the brightest star in the pentagon-shaped constellation Auriga and was the prized pet of the Charioteer. In Greek its name means "little she-goat."

Capella is a notorious twinkler. Although all stars twinkle, this one seems to attract special attention. When Capella is low in the sky, its light appears to change color, flicker, and dance dramatically. It can even flash alternately red, white, and blue. The effect is especially noticeable on October evenings when Capella is rising in the northeast. Astronomers are not sure what causes Capella to have such an especially sparkling personality, but it may be due to the fact that the light you see doesn't come from just one star. Capella is a multiple star system with four stars orbiting each other—two golden suns and two little red ones. The Capella system lies about 43 light-years from Earth.

### HOW TO FIND IT

**1** Capella and Sirius are the stars at the pointy ends of the Winter Football and are also the brightest stars in this expansive star pattern. During the heart of winter, if you lie back in the grass (or snow) and look straight up, Capella will greet you. When it is this high in the sky, it will not twinkle as much, but you can't mistake its dazzling appearance.

**2** The guide stars from Orion that brought you to Auriga can also take you to Capella. Start with the middle star in Orion's Belt, Alnilam, and draw a line from there through Orion's head star, Meissa. Continue that line of sight past Meissa and travel about 36 degrees. That will take you right to Capella.

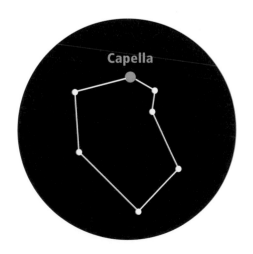

"When you're on Earth, if you go to the top of a mountain, the stars look much brighter than they do at sea level. And because the space shuttle is above Earth's atmosphere, it's like being on a very, very high mountain. So they look brighter, but not bigger."

Sally Ride, astronaut and physicist

# The Spring Sky

Each change of season provides an entire flock of new stars and constellations to learn. As March and April arrive, the stars of winter are still visible, but instead of being in the eastern half of the sky they are now positioned more toward the west. A new bevy of spring constellations has arrived to take their place and inhabit the eastern sky.

The spring sky is definitely not as dramatic and star-studded as the winter sky. While you have the Winter Football and so many bright stars visible in the winter sky, there are only three first magnitude stars among all the spring constellations: Regulus, Arcturus, and Spica. Together they form a gigantic star pattern called the Spring Triangle that makes a terrific jumping-off point to the other fainter and subtler constellations of spring.

In this section we will take a closer look at the three constellations that make up the Spring Triangle—Leo, the Lion; Boötes, the Bear Driver; and Virgo, the Maiden—as well as the major stars that lie within these three constellations. Then we will check out the largest constellation in the entire sky, Hydra, the Many-Headed Snake, which stretches almost across the entire southern sky during the heart of spring. We'll also look at the smaller constellations that ride on its back—Corvus, the Crow, and Crater, the Cup.

We'll continue our tour of the spring sky by identifying several small and subtle star patterns in Coma Berenices, Berenice's Hair; Corona Borealis, the Northern Crown; and Cancer, the Crab. Finally we'll meet a sprawling constellation that may have muscle but isn't very bright: Hercules. We will tie the legendary strongman into the untimely deaths of three other springtime constellations (Leo, Hydra, and Cancer) and whose appearance in the eastern sky tells us that summer will be coming soon.

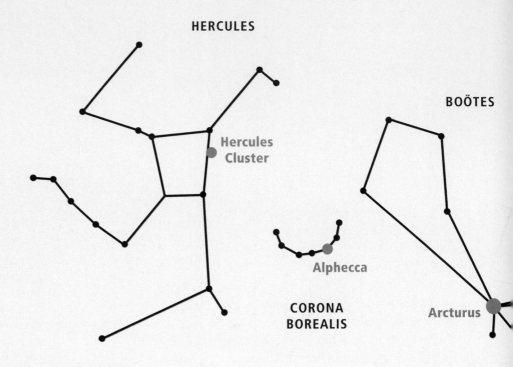

HERCULES

BOÖTES

Hercules
Cluster

Alphecca

CORONA
BOREALIS

Arcturus

## Spring Sky Constellations

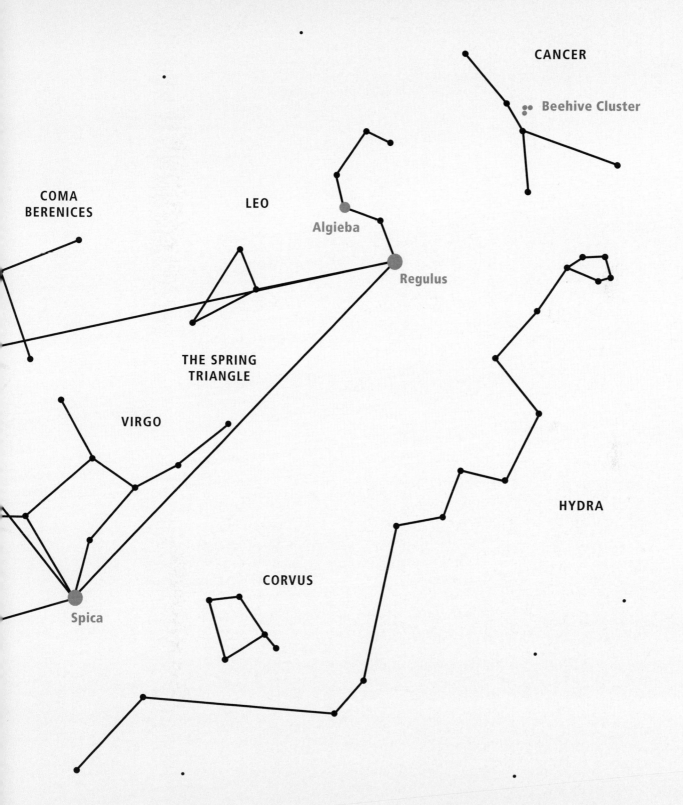

CANCER

Beehive Cluster

COMA
BERENICES

LEO

Algieba

Regulus

THE SPRING
TRIANGLE

VIRGO

HYDRA

CORVUS

Spica

# 38

# THE SPRING TRIANGLE

CLASSIFICATION: **ASTERISM** | VISIBILITY: **EASY**

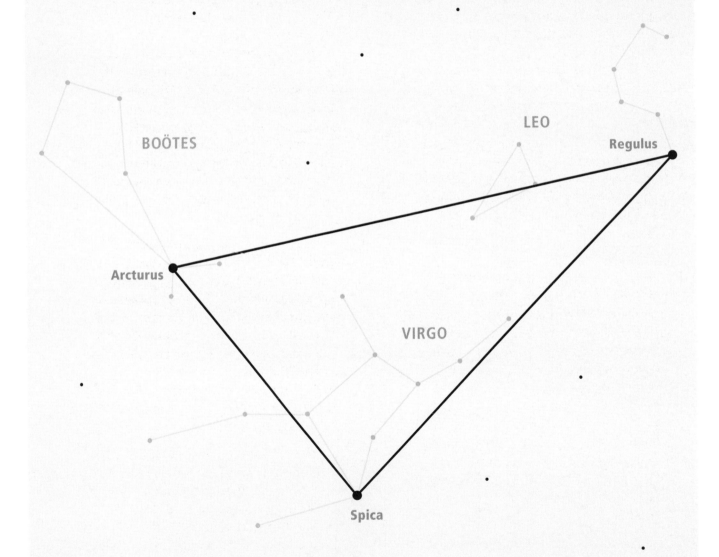

BOÖTES

LEO

Regulus

Arcturus

VIRGO

Spica

If you connect the dots from the three brightest stars in the spring sky, you can form a giant triangle. Known as the Spring Triangle, this unofficial constellation, or asterism, is 60 degrees long and 30 degrees wide, and it covers most of the southeastern sky during early spring and much of the southwestern sky in late spring. It is huge and is super easy to identify since these are the only bright stars in this section of the sky.

The three stars that make up the corners of the Spring Triangle are named Arcturus, Spica, and Regulus. These stars can help you identify some rather indistinct star patterns around the spring sky. Arcturus is the brightest star in the cone-shaped constellation Boötes, the Bear Driver; the star Spica can show you where the hand of Virgo, the Maiden, resides; while Regulus shows you the heart of Leo, the Lion.

## HOW TO FIND IT

**1** The easiest way to find the Spring Triangle is to look for the stars at each of the triangle's three corners: Regulus, Spica, and Arcturus. There are no other bright stars in the spring sky, so you can easily identify them by their position within the giant star pattern. Regulus leads the way in this asterism and is the first star of the Spring Triangle to rise and the first one to set. You can start to see it shining its blue-white light in the eastern sky after dark in **MARCH**. It and the constellation Leo is your signal that **SPRING** is almost here.

**2** Arcturus is the next to pop up above the eastern horizon along with the constellation Boötes, followed closely by Spica and its constellation, Virgo. Arcturus will be farther to the left (or closer to the northeast), while Spica will be over to the right (or closer to the southeast). This is a giant triangle covering most of the eastern sky as **SPRING** breaks. The angular distance from Regulus to Spica is 54 degrees. Regulus to Arcturus is the longest side of the Spring Triangle at 59 degrees. And even the closest corner stars, Spica and Arcturus, are still 33 degrees apart. You can measure these angles with your fist at arm's length—remember each fist-width is equal to 10 degrees. So you can see that the Spring Triangle is huge!

**3** You can start to see the entire Spring Triangle after dark in **APRIL**. But the best months to see the complete asterism are during **MAY** and **JUNE** when it stretches high across the southern sky. By **JULY**, Regulus sets in the west and only the other two stars in the Spring Triangle remain. Spica is gone in **SEPTEMBER** and Arcturus lingers into the **OCTOBER** sky, outlasting the **SPRING** and **SUMMER** seasons.

**4** Sometimes the planets wander through to mess up the outline of the Spring Triangle. Both Regulus and Spica lie near the ecliptic—the pathway of the planets. So do not be surprised to find extra bright lights between Regulus and Spica from time to time. These planets may ruin the perfect triangle shape, but they make for great bonus objects to observe.

# 39

# LEO

## *The Lion*

| CLASSIFICATION: **CONSTELLATION** | VISIBILITY: **EASY** |

Megrez

Phecda

**BIG DIPPER**

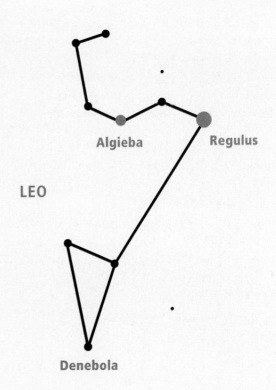

Algieba

Regulus

**LEO**

Denebola

There is no surer sign in the heavens that spring is back than seeing the constellation Leo, the Lion, in the sky. Leo is recognizable by the six stars shaped like a backward question mark—also called the sickle—which form its head. The bright star Regulus is the dot in the question mark and designates this King of the Beasts. The back end of Leo is marked by a triangle of stars, the farthest east being his tail, Denebola.

Ancient Greek mythology equated the constellation Leo with the Nemean Lion, whose hide was so tough that no sword or arrow could pierce it. This lion was menacing the countryside, killing villagers right and left. Hercules, the demigod and all-around he-man, was called in to take care of the situation as the first of his twelve labors (the twelve "impossible" jobs he was forced to accomplish in order to atone for killing his wife and kids during a fit of madness).

Instead of attacking Leo with a sword, Hercules wrapped his muscled arms around the beast's neck and strangled the lion to death. If you ever see pictures of Hercules later in his life, he's always wearing a lion skin. That's Leo's hide. It was like wearing bulletproof armor, and it protected Hercules from attacks. One question: How did Hercules cut Leo's hide to fashion his outfit if it was impenetrable? Answer: He used the lion's own claws to do the job.

## HOW TO FIND IT

**1** Leo is an easily recognized star pattern because of the lion's distinct head and rear end. Look for a group of stars that form a sickle shape or backward question mark. Leo's brightest star, Regulus, stands in as the dot at the bottom of the question mark that is supposed to be the lion's head and mane.

**2** In late **WINTER** look for Leo low in the east after dark. His head rises first and lies to the left of the Winter Football. On **MARCH** evenings you can spy the stars of the lion halfway up in the eastern sky. In mid-**SPRING** he's roaring high in the southern sky. And by the start of **SUMMER** Leo is over in the western sky.

**3** You can also use the bottom of the bowl of the Big Dipper's spoon to point you toward Leo's place in the sky. If you connect a line between the Big Dipper's faintest star, Megrez, and the bottom left star in the spoon, Phecda, and keep going about 40 degrees, that will take you to Leo's head.

## 40 | REGULUS

*The Little King*

CLASSIFICATION: **STAR** | VISIBILITY: **EASY**

If Leo, the Lion, is the king of the beasts, then he needs a royal star to go with him. And indeed, Regulus delivers. The name comes from Latin and means "little king." Regulus was one of the Four Royal Stars of ancient Persia, along with Fomalhaut (in Piscis Australis), Aldebaran (in Taurus), and Antares (in Scorpius). Each royal star reigned over a season of the year, and Regulus was the star whose departure from the sky signaled the end of spring and the coming of summer.

Regulus is blue-white in color and lies about 78 light-years from Earth. Astronomers have discovered that the light you see from Regulus actually comes from four stars that revolve around each other. The largest one, called Regulus A, is almost 4 times more massive than our Sun. Regulus A also spins rapidly, causing its gases to accumulate more around its equator. Astronomers characterize Regulus's shape as an oblate spheroid. You can liken it to a piece of M&M's candy. This is the star you are seeing in the spring sky with your naked eye, while the other three stars in the system require huge telescopes to spot.

### HOW TO FIND IT

**1** Leo's head stars look like a sickle shape or backward question mark. Regulus can be found at the handle of the sickle or, if you prefer, the bright dot at the bottom of the backward question mark.

**2** Look for this formation of stars in the eastern sky in early **SPRING**, high overhead in late **SPRING**, and over in the western sky as **SUMMER** begins. Regulus stands out as the brightest star in the constellation and in its quadrant of the sky.

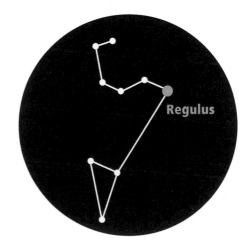

# ALGIEBA

## *The Double*

| CLASSIFICATION: **STAR** | VISIBILITY: **MODERATE** |

Leo's second brightest star is Algieba. In a rough translation from Arabic the name means "the forehead." However, since this star is so low on the lion's head, most people refer to Algieba as the lion's mane.

Algieba is about 130 light-years away from Earth and may look a little yellow or orange to the naked eye. If you're viewing from a dark sky, you may notice a faint star just below Algieba. This star is called 40 Leonis, and although it looks like it goes with Algieba, the two are unrelated. At about 60 light-years closer to us, 40 Leonis just happens to be in the same line of sight with distant Algieba. However, if you look at Algieba through a telescope and magnify it more than 100 times, you'll discover Algieba's secret twin. Algieba is actually a double star—two stars that revolve around each other. When you put your eye up to a good telescope, they look like two golden jewels that are so close together they almost seem to touch.

Although the two stars are often lumped together under the name Algieba (since you can only see the brighter of the two with the naked eye), astronomers refer to them as Gamma1 Leonis and Gamma2 Leonis. Since this pair of stars makes a handsome sight, astronomers have taken an even closer look at this system with some of the best telescopes in the world, and they discovered that Gamma1 Leonis (the brighter star) has at least one planet circling it. That means if you lived on that planet, you'd have two suns in your sky.

### HOW TO FIND IT

**1** To find Algieba you must first look for the sickle shape of stars that form the head of the constellation Leo, the Lion. If the brightest star in Leo, Regulus, is the handle of the sickle, then Algieba, Leo's second brightest star, is the bottom star of the blade.

**2** Look for Leo's head and head stars in the eastern sky in early **SPRING**, high overhead in late **SPRING**, and over in the western sky as **SUMMER** begins.

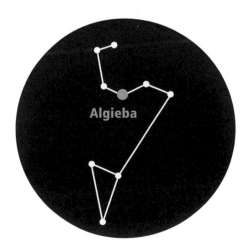

# BOÖTES

*The Bear Driver*

| CLASSIFICATION: **CONSTELLATION** | VISIBILITY: **MODERATE** |

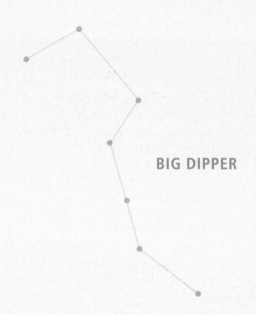

BIG DIPPER

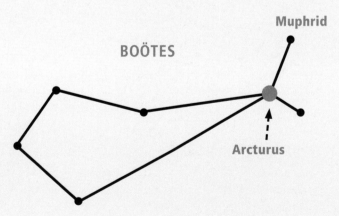

Muphrid

BOÖTES

Arcturus

Boötes (pronounced *Bo-OH-teaze*) is a kite-shaped constellation that first appears in the spring sky looking like a wide tie hanging from the invisible neck of an invisible businessman.

However, Boötes is known as the Bear Driver, since he can be found chasing after the big bear constellation, Ursa Major. The ancient Greeks noticed that the stars in Boötes just happened to follow Ursa Major in their nightly journey around the North Star. Just why Boötes was chasing the bear was unclear, but to add more credence to the story there is also another minor and very faint constellation between Boötes and Ursa Major called Canes Venatici. The stars in this constellation are supposed to be Boötes's two dogs that are helping him chase the bear. You'll need a very dark sky (and tons of imagination) to see Canes Venatici, however.

As the night goes on, Ursa Major circles the North Star and climbs higher in the sky, while Boötes (and his dogs) follow. During the summer months you can find Boötes very high in the southern sky, and even through mid-October you can spot him in the western sky.

An ancient Greek myth portrays Boötes as an inventive man who overcomes extreme adversity. One day Boötes was walking through the woods when he was robbed by his brother. When he stumbled home, battered and penniless, he discovered that his good-for-nothing brother had taken possession of everything he owned, including his house and land. Undaunted, Boötes found a new place to live and invented a plow that could be pulled by oxen. This story also fits in extremely well with the old English view of the Big Dipper as the Plough. Boötes can be seen right behind the plow, pushing it around the sky as he circles the North Star.

## HOW TO FIND IT

**1** The easiest way to find Boötes is to use the more recognizable stars in the Big Dipper to point the way. First find the Big Dipper and go to the three stars that form its curved handle. Connect the dots on those three stars and continue the arc that they create. If you keep going for another 30 degrees, you will arc your way to Arcturus. In fact, one of the catchiest sayings in stargazing is "Follow the Arc to Arcturus."

**2** Boötes hangs high in the southern sky in **MAY** and **JUNE** and keeps chasing the big bear toward the west in **JULY** and **AUGUST**. You will see him for the last time in **SEPTEMBER** and early **OCTOBER** when the constellation's outline looks more like an outline of a kite that is standing upon Arcturus.

## 43 | ARCTURUS

### *The Bright*

CLASSIFICATION: **STAR** | VISIBILITY: **EASY**

The name Arcturus comes from Latin and means "guardian of the bear." The name is apt since the star lies relatively close to the major constellation Ursa Major. Polynesian astronomers call it Hōkūle'a, meaning the "star of joy," and used it as a guide star to help them navigate around the Pacific Ocean. Since Arcturus can appear straight overhead from the Hawaiian Islands, Polynesian sailors used its elevation to determine their latitude. When Arcturus was overhead, they sailed east or west until they ran into the islands.

As the fourth brightest star in the entire sky, Arcturus's intense brightness will definitely attract your attention. When you first see it, you may wonder if it is a plane or UFO. Only Sirius (in the winter sky) and Canopus and Alpha Centauri (two stars in the southern sky that are not visible from most places in the United States) are brighter.

Even with the naked eye you will notice that Arcturus shines with an orangish glow. This color tells astronomers that Arcturus has a surface temperature of about 7,300 degrees Fahrenheit (a much cooler star than our 10,000-degree yellow Sun). Arcturus is about 36.7 light-years from Earth, making it one of the closer stars. It is an orange giant star with a diameter 25 times that of the Sun, and it shines 170 times brighter.

The vivid light of Arcturus famously triggered the start of the 1933 World's Fair in Chicago. When this star's light fell on a sensor, it activated a switch that illuminated the fairgrounds on opening night.

### HOW TO FIND IT

**1** To find Arcturus, first find the Big Dipper and trace out the three stars that form its curved handle.

**2** Continue the arc of those stars away from the Big Dipper for 30 degrees until you run into Arcturus. If you "follow the arc to Arcturus," then you will find the brightest star in the constellation Boötes.

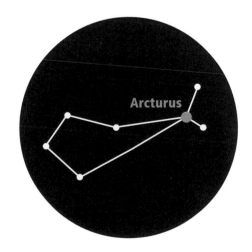

"Space is big. Really big. You just won't believe how vastly, hugely, mind-bogglingly big it is. I mean, you may think it's a long way down the road to the chemist, but that's just peanuts to space."

Douglas Adams, author and humorist

# 44

# COMA BERENICES

*Berenice's Hair*

| CLASSIFICATION: **CONSTELLATION** | VISIBILITY: **DIFFICULT** |
| --- | --- |

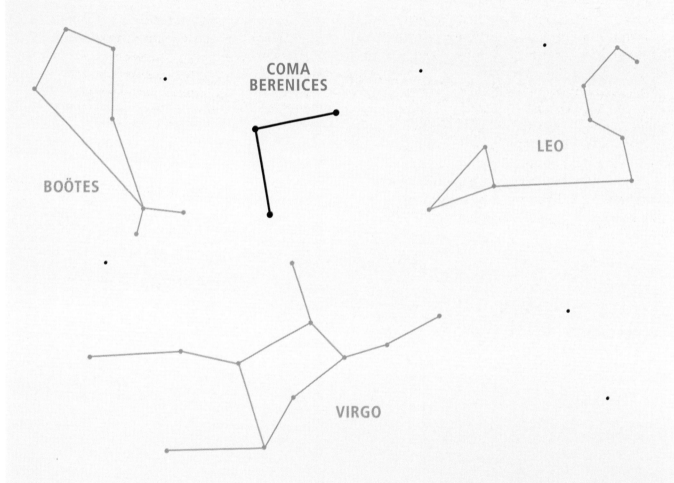

COMA
BERENICES

BOÖTES

LEO

VIRGO

The faint smattering of stars in the spring sky called Coma Berenices has a hair-raising tale. In Greek mythology Berenice was the beautiful queen of Egypt known for her flowing tresses. When her husband went off to war, Berenice asked Aphrodite to protect her beloved in battle. In return, if he returned to her safely, she would cut off her long hair as a gift to the goddess. When the king returned unharmed to her side, Berenice stayed true to her word and lopped off her hair. The hair was placed in the temple where it mysteriously disappeared. Who dared to take the queen's beautiful hair?

Heads were going to roll if the culprit was found. Luckily a court astronomer came to the rescue—he found the missing locks up in the heavens. The hair was such a pleasing sacrifice to Aphrodite that she took it and placed it in the sky for all to see. And so the glory of Berenice's hair reached new heights and lives on in the stars.

## HOW TO FIND IT

**1** Coma Berenices is incredibly faint and can only be detected from a dark sky. On some star charts astronomers outline the three brightest stars in the constellation with a right-angle shape, but this doesn't capture the true nature of the entire star pattern.

**2** At first glance it will look like a diffuse cloud about 20 degrees above Virgo and about halfway between the constellations Boötes and Leo. When you squint in this region of space, you may be able to make out a swarm of individual stars that evoke an image of flowing hair.

# 45

# HYDRA

## *The Many-Headed Snake*

CLASSIFICATION: **CONSTELLATION** | VISIBILITY: **DIFFICULT**

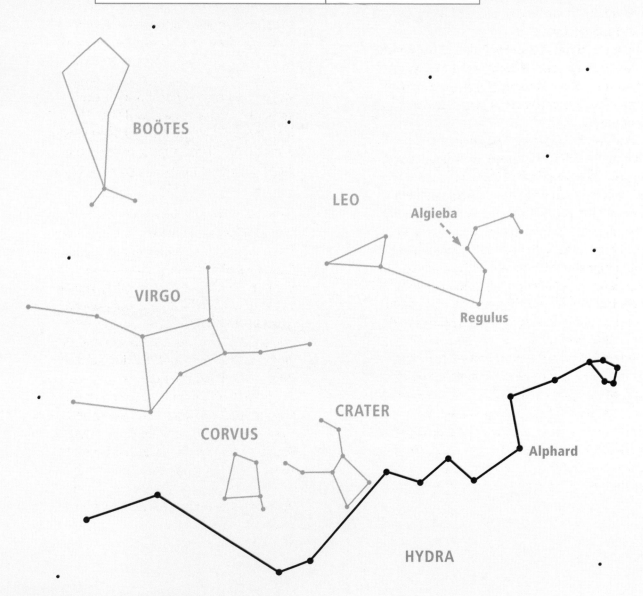

BOÖTES

LEO

Algieba

VIRGO

Regulus

CRATER

CORVUS

Alphard

HYDRA

Hydra is the largest constellation and coils across most of the southern sky every April and May evening. From heads to tail Hydra spans more than 100 degrees. Remember that your fist at arm's length equals about 10 degrees. So Hydra is ten fist-widths long—which is more than one-quarter of the way around the horizon!

Hydra also has company. Riding on his back are two fainter constellations called Corvus and Crater. Corvus is definitely the brighter of the two and is a crafty crow, while Crater is a cup of water.

Like the constellation Leo, Hydra was a victim of Hercules, the Greek warrior and he-man. After strangling the Nemean Lion, Hercules had to kill the seven-headed Hydra of Lerna. This was the second of his twelve labors, and it was not an easy one to accomplish. The Greeks believed that Hydra was a terrible monster with seven, eight, or even nine heads. People had tried to kill Hydra before but discovered that every time they cut off one of the heads, two more would grow back in its place!

Hercules went to slay the monster with help from his cousin Iolaus. Whenever Hercules cut off a head, he would signal Iolaus to come over with a hot iron and cauterize the wound shut so that no heads could grow back. It worked! After the seventh head fell, the Hydra was dead and joined Leo, the Lion, in Hercules's monstrous pet cemetery in the sky.

## HOW TO FIND IT

**1** To find Hydra, first find the constellation Leo, the Lion. Leo usually stands higher in the sky than Hydra, and his sickle of stars is much more distinct. Connect the dots of Leo's two brightest stars in the sickle, Algieba and Regulus, and continue that line of sight downward another 23 degrees. You will run right into Hydra's one bright star, Alphard, which means "the backbone of the serpent." Hydra's heads will look like a little ring of five faint stars at the westernmost (or farthest right) part of the serpent, and its tail will wind over to the left.

**2** Note that, although Hydra is huge, it can be a challenge to locate if the night sky has a lot of light pollution. But if you look carefully, Alphard should appear to be a little redder than the other stars in the spring sky. If you are viewing from a small city or out in the country, then you should have no trouble making out Hydra's distinctive and multifaceted heads. The head stars can set long before the heart and tail of Hydra. The ring of stars marking Hydra's head are prominent from **FEBRUARY** to **MAY** while Alphard, the backbone of the serpent, shines from its solitary position from **MARCH** until **JUNE**. You can see the entire Hydra in the evening hours only during the month of **MAY**.

# 46 | CORVUS

*The Crow*

| CLASSIFICATION: **CONSTELLATION** | VISIBILITY: **MODERATE** |

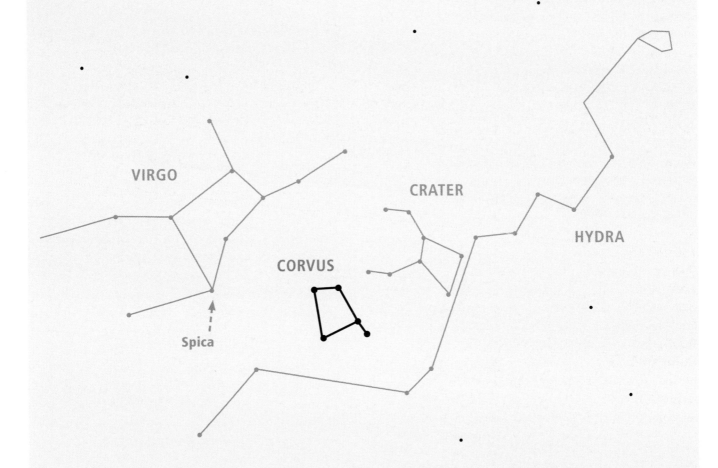

VIRGO

CRATER

CORVUS

HYDRA

Spica

In Greek mythology, one day Apollo worked up a mighty thirst. Apollo was the god of just about everything—from music to truth, and light to healing. But he was best known for guiding the celestial horses and chariot that carried the Sun daily across the sky. Since Apollo barely had any free time to himself, he often employed a crow named Corvus to act as a personal assistant.

When Apollo grew thirsty from carrying the Sun around all day, he sent Corvus with his favorite cup to fetch water from the river.

Several cultures, including the Greeks, looked upon crows as intelligent and crafty creatures. Corvus was no exception. It was such a beautiful sunny day (thanks to the work of Apollo) that Corvus stopped along the way to eat some juicy figs and take a little nap. When he woke up, the day was almost over and the Sun had almost set, and he knew he was going to feel Apollo's wrath for being so late to return. Corvus crafted what he thought was a devious excuse. He filled up the cup with water, picked up a water snake from the river, and then flew back to Apollo. Corvus explained, "Sorry I'm late, Apollo. But this water snake wouldn't let me get any water. So I brought him here to show you I wasn't lying." Apollo, seeing through the horrible falsehood, threw the crow, cup, and snake into the sky to become the constellations Corvus, the Crow; Crater, the Cup (a much fainter nearby constellation); and Hydra, the Many-Headed Snake. Now, Corvus must ride on the snake's back for eternity with the cup of water just out of his beak's grasp, making him forever thirsty. This legend was said to explain why earthly crows are cursed with such rough, raspy voices.

## HOW TO FIND IT

**1** Although Corvus is a small constellation it is moderately easy to find in the sky. Look for a trapezoid shape of four semi-bright stars riding on the back of Hydra.

**2** Corvus, the Crow, is closer to the tail-end of Hydra and is only a short distance away from Spica, the bright blue star that lies in the constellation Virgo, the Maiden.

**3** In fact, in **APRIL** you can see Spica rise with Corvus in the southeastern sky. Spica is about 17 degrees to the left of Corvus. In **MAY** the trapezoid shape of Corvus stands about 30 degrees above the southern horizon. By **JUNE** and early **JULY**, Corvus, the Crow, will be angled down lower in the southwestern sky and Spica will be 17 degrees above Corvus.

# 47

# VIRGO

## *The Maiden*

| CLASSIFICATION: **CONSTELLATION** | VISIBILITY: **MODERATE** |

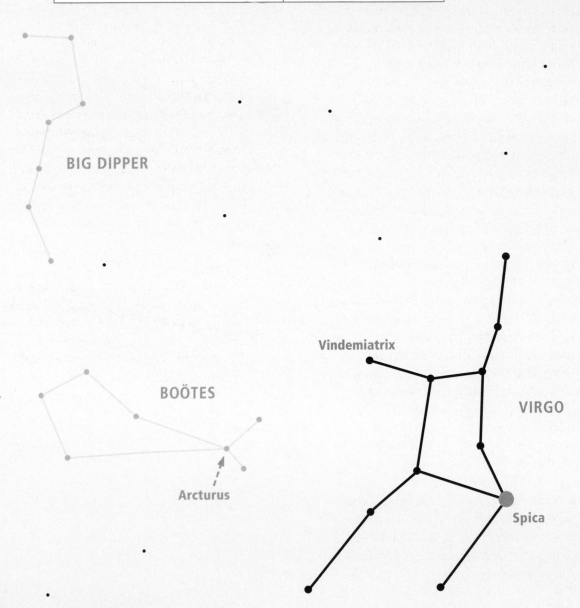

BIG DIPPER

BOÖTES

Arcturus

Vindemiatrix

VIRGO

Spica

Virgo is known as the virgin or maiden. It is a huge constellation (only Hydra is larger) that appears in the southeastern sky as spring begins. Illustrations of her in the sky show her reclining serenely while holding an ear of wheat in one hand and a dove of peace in the other. However, it is difficult to imagine Virgo's silhouette when looking at the night sky. Many of her stars are dim, and the figure really doesn't look much like a maiden, but she is still moderately easy to find if you can identify her brightest star, Spica (pronounced *SPY-kah*).

The ancient Greeks likened Virgo to the daughter of their chief god, Zeus, and Themis, the goddess of justice. Virgo lies in the sky next to another zodiac sign, Libra, the Scales. This location displays her association with the scales of justice. Unfortunately, Virgo is not holding the scales since the constellation Libra lies at the maiden's feet.

In ancient Egyptian mythology, Virgo was associated with Isis, the supreme mother goddess. As legend has it, one day Isis was eating corn in the sky when a monster of enormous size and viciousness named Typhon came upon her and chased her far and wide. As Isis fled from the great beast, she dropped pieces of corn all across her path. These kernels of corn turned into stars and became the Milky Way.

In India, Hindu astronomers believed Virgo to be Kanya, the maiden and mother of the great Krishna. Arab astronomers initially included Virgo in a giant Lion constellation that covered a huge swath of the spring sky, while others in the Middle East called the stars of Virgo the Barking Dogs. Over time these astronomers changed this constellation to conform to the Greek myths and, toward the end of the first millennium A.D., began to call Virgo by the name Al Adhra al Nathifah, which translates to "the Innocent Maiden."

## HOW TO FIND IT

1 To locate Virgo's resting place in the sky, start with the Big Dipper. Follow the stars in the curved handle of the Big Dipper and continue that arc for 30 degrees until you reach the dazzling orange star, Arcturus, the brightest star in the constellation Boötes (remember to "Follow the Arc to Arcturus"). After you find Arcturus, straighten out the arc and keep going for another 30 degrees until you run into bright, blue Spica. The full saying is: "Follow the Arc to Arcturus. Then hit a Spike to Spica!"

2 Virgo rises in the east-southeast after sunset in late-**MARCH** and early **APRIL**. At that position her outline appears to be standing above the horizon with one arm up in a disco-like pose. During **MAY** and **JUNE** evenings, you can find Virgo about halfway up in the southern sky. And in **JULY** and **AUGUST** she looks as if she is falling headlong into the southwestern sky.

# 48

# SPICA

## *The Blue-White Diamond*

| CLASSIFICATION: **STAR** | VISIBILITY: **EASY** |
|---|---|

Spica is the real star of Virgo. It is a brilliant blue-white in color and lies 260 light-years away. It has been called the Queen Star of the Spring and the Star of Prosperity. However, in Latin, *Spica* literally means "ear of wheat," which Virgo is allegedly holding in her left hand.

The Egyptians had a special affinity for Spica because of its location along the zodiac, and they built many temples to its movement. At key times in the year, the light of Spica would penetrate through deep shafts into these temples.

Spica lies near the celestial equator which means it is prominently visible every night for about 6 months (March–September). Its rising at sunset marked the official beginning of the spring planting season for cultures across the Northern Hemisphere. The setting of Spica at sunset signaled the fall harvest time.

Spica is also near the ecliptic—the apparent pathway that the Sun, Moon, and planets wander. About once a month, the Moon cozies up to Spica and can even cover it up (this is called an occultation). Planets are also frequent visitors to Virgo and Spica, and it is not unusual to see Venus or Mars nearby.

## HOW TO FIND IT

**1** Spica is the brightest star in Virgo and the only one in the constellation that stands out. First locate the Big Dipper and the three stars on its curved handle. If you continue that arc from the Big Dipper and go another 30 degrees you will find Arcturus in the constellation Boötes.

**2** Then straighten out the arc and keep going for another 30 degrees until you run into bright, blue Spica. Remember the full saying and you won't get lost: "Follow the Arc to Arcturus. Then hit a Spike to Spica!"

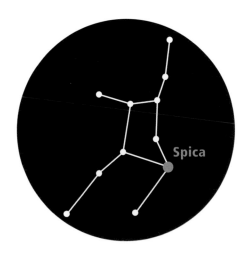

"We are not simply in the universe, we are part of it. We are born from it. One might even say we have been empowered by the universe to figure itself out—and we have only just begun."

Neil deGrasse Tyson, astrophysicist

# 49

# CORONA BOREALIS

*The Northern Crown*

| CLASSIFICATION: **CONSTELLATION** | VISIBILITY: **MODERATE** |

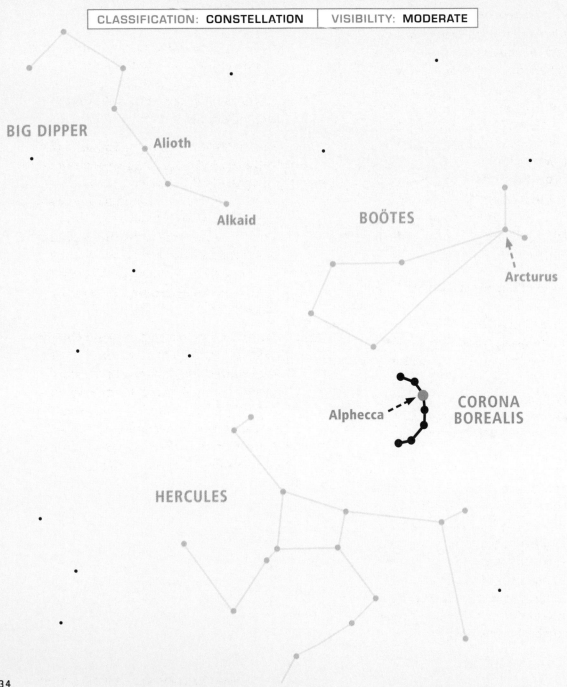

BIG DIPPER

Alioth

Alkaid

BOÖTES

Arcturus

Alphecca

CORONA
BOREALIS

HERCULES

Corona Borealis, a.k.a. the Northern Crown, is one of the most identifiable constellations in the sky. Although small and dim, once you find the outline of the seven stars that form a subtle semicircle of sparkling jewels, you will always remember it.

The Greeks likened this star picture to the crown presented to a beautiful maiden named Ariadne who fell in love with Theseus, the prince of Athens. Unfortunately for their love affair, Theseus was chosen to be sacrificed in the great labyrinth of King Minos whose twisting halls were roamed by the vicious Minotaur— a creature that was half man, half bull. Before Theseus was thrown into the labyrinth, Ariadne gave him a sword to kill the beast and a huge spool of thread. Theseus tied one end of the thread to the entrance and reeled out the thread as he walked through the maze. This way, he would be able to find his way out. Theseus succeeded in slaying the Minotaur and followed the thread back to Ariadne's arms.

The couple lived happily for only a few years, then Theseus grew bored with family life and left Ariadne for her sister, Phaedra. The god of wine, Dionysus, took pity on Ariadne for this act of treachery and granted her the most beautiful crown in the world. Upon her death the crown was placed in the skies for all to see.

This formation of stars is so distinct that many other cultures envisioned an image in this region of the sky. For example, the Australian Aborigines thought this constellation was a boomerang flying through the heavens. In one Native American legend it is the cave where the great bear, Ursa Major, lives. And at different seasons, depending on how it's tipped, this constellation can look like a smile or a frown, an umbrella or a bowl.

## HOW TO FIND IT

1. Corona Borealis resides between Boötes, the Bear Driver, and the mighty Hercules. One way to find the crown is to first find the Big Dipper's handle. This time, however, connect the dots of the first and third stars in the handle (Alioth and Alkaid) and continue that line from these two stars outward from the spoon for another 30 degrees. That will take you to Corona Borealis.

2. Corona Borealis rises after Boötes in the northeast, so you probably won't notice this constellation until **APRIL**. By **JUNE** and **JULY** it rides so high in the southern sky that you'll want to lie back in the grass and look almost straight overhead to find it. The Northern Crown can still be observed every evening into early **NOVEMBER** when it sets in the northwestern sky.

3. This constellation is not incredibly bright so do not expect it to stand out like Orion's Belt or shine as bright as the stars in the **SPRING** Triangle. Corona Borealis is a small, subtle grouping of stars but once your eye alights on them, you will discover a semicircle of stellar gemstones like no other in the northern sky.

# 50 | ALPHECCA

*The Gem in the Crown*

| CLASSIFICATION: **STAR** | VISIBILITY: **MODERATE** |

The Shawnee Indians of the Ohio River Valley saw the seven stars of Corona Borealis, the Northern Crown, as seven beautiful maidens. One story tells of a hunter named White Hawk who was sleeping in a clearing one warm summer day. He was jolted awake by a giant silver basket descending by a silver cord from the clouds. Upon closer inspection, White Hawk noticed that the basket was occupied by seven stunning young women. When the basket touched the ground, the maidens stepped out of the basket and danced wildly in a ring.

White Hawk watched from a good distance but crept closer. He was so enthralled that he wanted to ask the most beautiful maiden to be his bride. But when he approached, the women jumped into the basket, pulled the silver cord, and flew back into the sky once more.

The next day White Hawk returned to the same clearing at the same time. This time he disguised himself as a rabbit in the hopes of getting closer to the young ladies. Sure enough, the basket with the women descended again to the ground. They danced. He hopped closer (as a bunny), but the disguise didn't work. The women jumped back into the basket, pulled the cord, and flew skyward.

So he tried it again the next day—this time disguised as a mouse. For some strange reason the disguise worked. He scurried up to the seven maidens, grabbed the fairest of them all, and carried her away.

White Hawk was said to be represented by Arcturus, the brightest star in the constellation Boötes, while the fairest of the maidens became the brightest star in the Northern Crown, Alphecca. The name Alphecca is not Shawnee in origin but instead comes from an Arabic word meaning "bright one of the dish."

## HOW TO FIND IT

**1** In order to locate Alphecca, first find the Big Dipper. Trace a line from the handle star Alioth to Alkaid.

**2** Keep going. After traveling another 30 degrees of sky, this line of sight will take you right to Alphecca, a star that is 75 light-years away and shines with a stark white light.

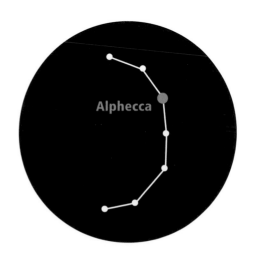

"I have loved the stars too fondly to be fearful of the night."

Sarah Williams, poet

# CANCER

## *The Crab*

| CLASSIFICATION: **CONSTELLATION** | VISIBILITY: **DIFFICULT** |

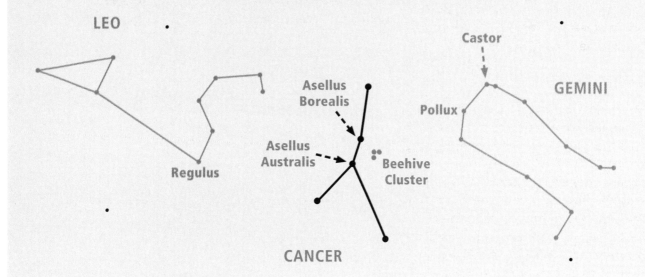

LEO

Regulus

Asellus
Borealis

Asellus
Australis

Beehive
Cluster

CANCER

Castor

Pollux

GEMINI

Cancer is a small and faint zodiac constellation visible in the spring sky. This star pattern is supposed to represent the killer crab sent by the goddess Hera (Zeus's rightfully jealous wife) to harass Hercules, the strongest guy in ancient Greece (and the illegitimate son of Zeus). The plan was to have the crab distract the mighty warrior during his battle with the seven-headed Hydra. However, before the crab could nip at his heels, Hercules made short work of it with one mighty step. Cancer, the Crab, was squished underfoot and vanquished to the sky to be honored along with Hercules's other conquests—Leo, the Lion, and Hydra, the Many-Headed Snake.

### HOW TO FIND IT

**1** The brightest part of Cancer, the Crab, is comprised of third and fourth magnitude stars that form an upside-down Y shape. They definitely will not stand out in your sky and may even be completely invisible if you're viewing from an urban location. The best way to find the area where Cancer resides is to use the more brilliant zodiac constellations on either side of it. Leo, the Lion, is to the left of Cancer, and the Gemini Twins are on the right. Identify the backward question mark for Leo's head and the dazzling head stars of Gemini, Pollux and Castor. Cancer sits about halfway between these two landmarks.

**2** Cancer rises earlier than any of the other **SPRING** constellations, so you can start to see it halfway up in the eastern sky in **FEBRUARY** and **MARCH**. During those months Pollux and Castor will be hanging 15 degrees above Cancer. The best month to see Cancer in the evening is **APRIL**, when the crab is almost straight overhead with Leo's brightest star, Regulus, 20 degrees to the left of it.

**3** By **MAY** and **JUNE**, Cancer is lower in the western sky, and it exits the evening sky altogether in **JULY**.

## 52 | THE BEEHIVE CLUSTER

### *A Swarm of Stars*

CLASSIFICATION: **STAR CLUSTER** | VISIBILITY: **DIFFICULT**

Within the constellation Cancer you can find a gorgeous group of stars. Some people say that the light from these stars resembles a swarm of bees circling a hive, and this grouping was therefore named the Beehive Cluster. This open star cluster consists of 1,000 stars with a common center of mass.

Known from antiquity, the Beehive Cluster inspired many astronomers and writers. Around the year 265 B.C. the Greek author Aratus described it as a "little mist." The Greek astronomer Hipparchus called the cluster a "little cloud" in 130 B.C. And about 200 years later the great astronomer Ptolemy, whose writing guided astronomers for the next 1,400 years, included it as one of the seven nebulas he could see with the naked eye. Some ancient Greeks and Romans also called this group of stars the Praesepe, or "the manger," with two donkeys—the stars named Asellus Borealis and Asellus Australis—resting among the stars nearby.

### HOW TO FIND IT

**1** The Beehive Cluster shines with an apparent brightness of a third or fourth magnitude star (much dimmer than the stars in the Big Dipper). So you will need a dark sky to see it clearly.

**2** If you can make out the upside-down Y shape of Cancer, then look for two stars at the junction of the crab's extremities. These are Asellus Borealis and Asellus Australis, the donkeys.

**3** Just to the right of the two donkey-stars, if you squint you might, just might, be able to see a small, faint, and fuzzy patch of sky. That is the Beehive Cluster. You may even be able to detect it much better if you don't look directly at it. Astronomers call this averted vision, when you allow your peripheral vision to see faint objects more clearly than if you stare straight at them.

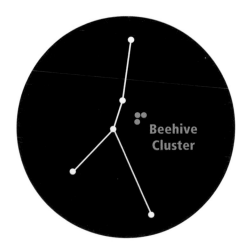

Beehive Cluster

"Everything was so new— the whole idea of going into space was new and daring. There were no textbooks, so we had to write them."

Katherine Johnson, mathematician for NASA

# 53 | HERCULES

*The Kneeler*

CLASSIFICATION: **CONSTELLATION** | VISIBILITY: **DIFFICULT**

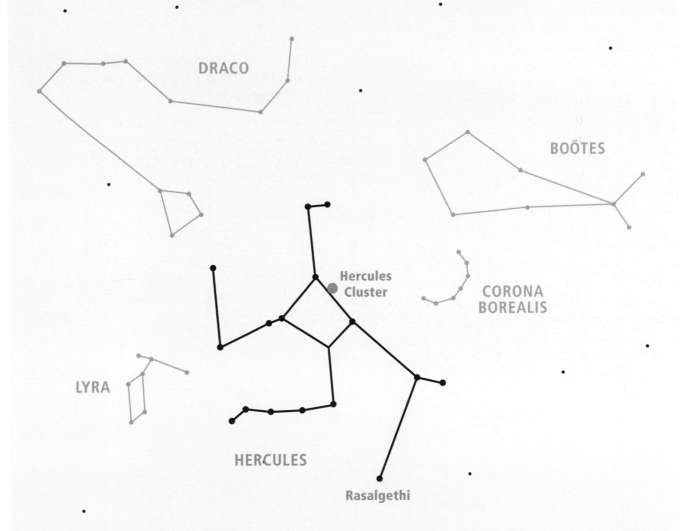

DRACO

BOÖTES

Hercules
Cluster

CORONA
BOREALIS

LYRA

HERCULES

Rasalgethi

In Greek mythology Hercules (or Heracles in Greek) was the son of Zeus and a mortal woman, Alcmene. His name was a slap in the face to Zeus's wife, Hera, since the name Heracles means "glory of Hera." Through a bit of sorcery, Hera drove Hercules stark raving mad—mad enough to kill his wife and kids. As punishment Hercules was forced to do twelve labors for King Eurystheus. At the completion of these labors, including the slaying of the Nemean Lion (the constellation Leo) and the many-headed serpent (the constellation Hydra), Hercules would achieve immortality.

All the stars that make up the constellation of this mythical demigod are dim. Only Hercules's head star, Rasalgethi, stands out, and even it barely shines brighter than third magnitude (only half as bright as the stars in the Big Dipper). Rasalgethi is an old Arabic word meaning "head of the kneeler." It has the uniquely red hue of a supergiant star. Rasalgethi's brightness also varies wildly—a common trait for red giants. When you find Hercules's head in the stars, aim a small telescope at Rasalgethi and you will discover that it is really two stars in one. Some ancient Greeks pictured Hercules in the stars kneeling with an upraised club in one hand and Hera's snakes in the other. Why he is upside down, no one knows.

## HOW TO FIND IT

**1** The best way to find the bulk of Hercules is to look for the keystone, which is a four-sided figure that makes up the torso of his body. Use the curving star pattern of Corona Borealis to point you to the keystone's location by following the lip of its semicircle of stars from right to left. At the end of the left side, exit the curve and keep going about 14 degrees farther. That will take you to the center of the keystone.

**2** Once you note the position of the keystone, you may be able to imagine Hercules's arms, kneeling legs, and head. His outline is totally spread out with legs and arms flailing in all directions.

**3** You can also identify Hercules's place in the sky by the more noticeable stars and constellations surrounding him. The four-sided head of Draco, the Dragon, generally flies above him, while Corona Borealis is to the right.

## 54 | THE HERCULES CLUSTER

*Totally Globular*

| CLASSIFICATION: **STAR CLUSTER** | VISIBILITY: **DIFFICULT** |

The most renowned globular cluster visible from the Northern Hemisphere is called M13, or the Hercules Cluster. It has about 300,000 stars in it and stretches across almost 170 light-years of the galaxy. But M13 is so far away (about 22,000 light-years) that you can just barely see it with the naked eye under a dark sky.

Globular clusters like M13 are much larger groupings than open clusters like the Seven Sisters, and they show very little open space between the stars within them. Through a small telescope they look like globes of light. But in a larger scope you can see the thousands of individual stars like a swarm of fireflies filling your entire view. When you observe a globular cluster you are seeing the light of hundreds of thousands of stars.

Globular clusters are some of the largest structures observed within the Milky Way, as well as some of the oldest. Within the Milky Way some globular clusters contain stars that are more than 11 billion years old. They may have been the first stars to form in the earliest stages of the Milky Way.

The *M* in M13 comes from French astronomer Charles Messier's last name. In the eighteenth century he catalogued 110 deep space objects like star clusters, galaxies, and nebulas. As he noted them on his charts, he numbered them M1, M2, M3, and so on. M13 was Messier's 13th find with his telescope.

### HOW TO FIND IT

**1** To find the Hercules Cluster, you must first find Hercules using the instructions on the previous page.

**2** After you identify Hercules, locate the two stars that make up the longest side of the keystone. If you start at the fainter of the two stars and head toward the brighter one, M13 is about one-third of the way along that path.

**3** Keep in mind that M13 is perhaps the faintest object described in this book and can only be detected with the naked eye under ideal viewing conditions. That said, it makes a terrific target for binoculars and can present a truly glorious spectacle through a telescope.

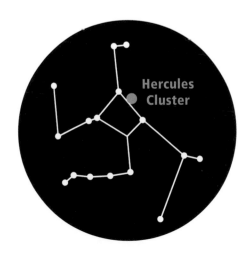

"Man will not always stay on Earth; the pursuit of light and space will lead him to penetrate the bounds of the atmosphere, timidly at first, but in the end to conquer the whole of solar space."

Konstantin Tsiolkovsky, one of the founders of modern astronautics

# The Summer Sky

Warm summer evenings are perfect for lying in the grass and looking up. Most of the brightest stars of the season are high overhead. The temperatures are perfect. And there's something amazing about that time when the Sun has set and the stars are slowly emerging from the royal blue sky. In this season it really seems that stargazing is a moment to slow down, breathe, think, and prepare.

There are so many stars in the summer sky that they may seem to be alive. Cygnus, the Swan, actually looks like a swan with a long neck, beak, and outstretched, starry wings. The tiny constellation Delphinus, the Dolphin, leaps out of the cosmic ocean, back arched, with the Milky Way sparkling above. Scorpius, the Scorpion, with its red beating heart star, Antares, crawls above the treetops.

Summer is the season to become a stargazer. Whether you're a kid or an adult, this is the time to lie back in the grass, get out of town, take a vacation to the countryside to experience a truly dark sky, view the stars of summer, and share the universe with those you love. So let's take a look at what you can see in the warm summer months of June, July, and August.

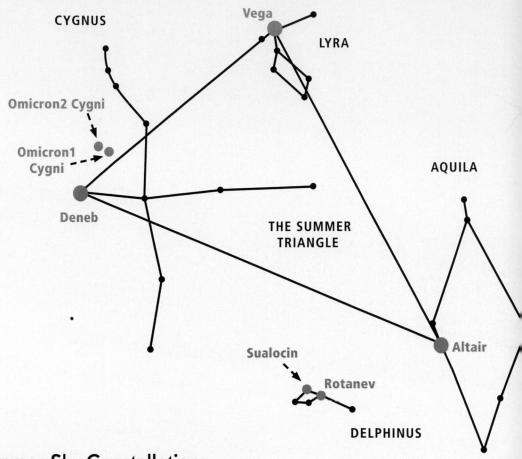

CYGNUS

Vega

LYRA

Omicron2 Cygni

Omicron1 Cygni

AQUILA

Deneb

THE SUMMER TRIANGLE

Sualocin

Altair

Rotanev

DELPHINUS

## Summer Sky Constellations

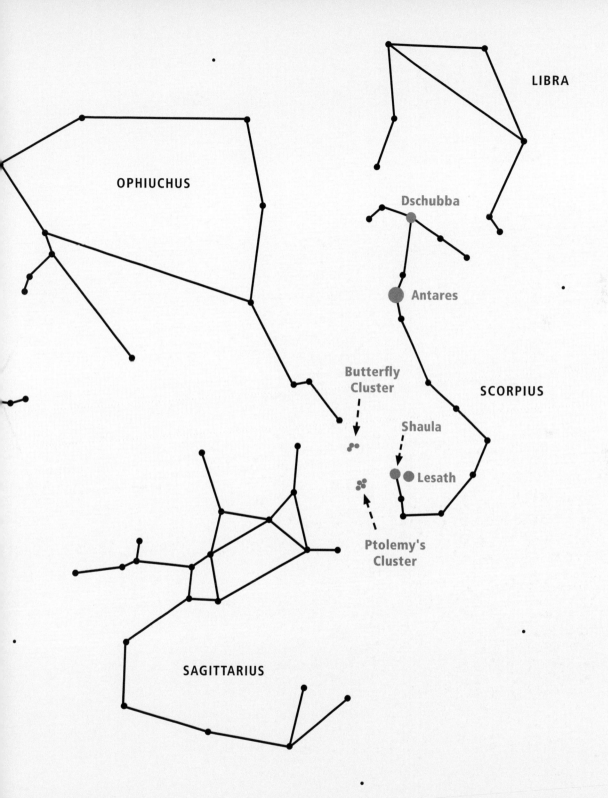

LIBRA

OPHIUCHUS

Dschubba

Antares

Butterfly
Cluster

SCORPIUS

Shaula

Lesath

Ptolemy's
Cluster

SAGITTARIUS

# 55

# THE SUMMER TRIANGLE

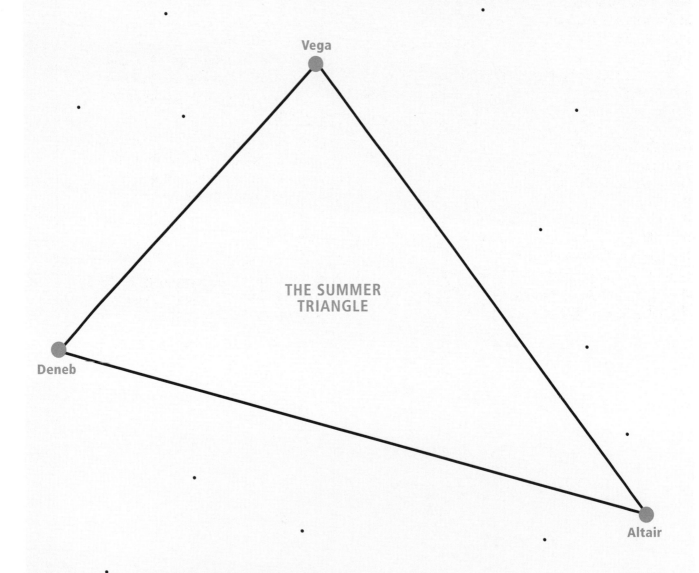

Vega

THE SUMMER
TRIANGLE

Deneb

Altair

In June, July, and August when you point your feet to the east and lie back you'll find a triangle of three celestial dazzlers—stars named Vega, Deneb, and Altair. When you connect these stars with lines, they form a large triangle measuring about 30 degrees long and 20 degrees high. Remember the easy way to measure degrees in the sky? Your fist at arm's length is about 10 degrees. So this shape in the sky will be two fists high and three fists long.

Astronomers call this formation the Summer Triangle, which, like the Big Dipper and the Spring Triangle, is an asterism (a recognizable shape of stars) rather than an official constellation. In fact, each star in the triangle is part of its own constellation. Vega is the brightest star in the constellation Lyra, the Harp. Deneb is the tail star of Cygnus, the Swan. And Altair is the eagle eye of the constellation Aquila, the Eagle. The Summer Triangle is super easy to identify, and you'll see all three stars every night in the summertime, even if you live in the heart of a city or under any light-polluted sky.

## HOW TO FIND IT

1. It is called the Summer Triangle because the three stars, Vega, Altair, and Deneb, become really noticeable at the start of the **SUMMER** season. On **JUNE** and **JULY** evenings you can find it just above the eastern horizon.

2. Vega leads the way and is highest in the sky during these months. Deneb rises next in the northeast while Altair, last but not least, trails behind the other two and sneaks above the treetops in the southeast.

3. In **AUGUST** and **SEPTEMBER** the Summer Triangle is almost straight overhead, so it is best observed by lying back in the grass with your feet pointed to the south.

4. You can still see it in **OCTOBER**, **NOVEMBER**, and even **DECEMBER** in the western sky after dark. It may rise up in the summertime, but the Summer Triangle spans several seasons.

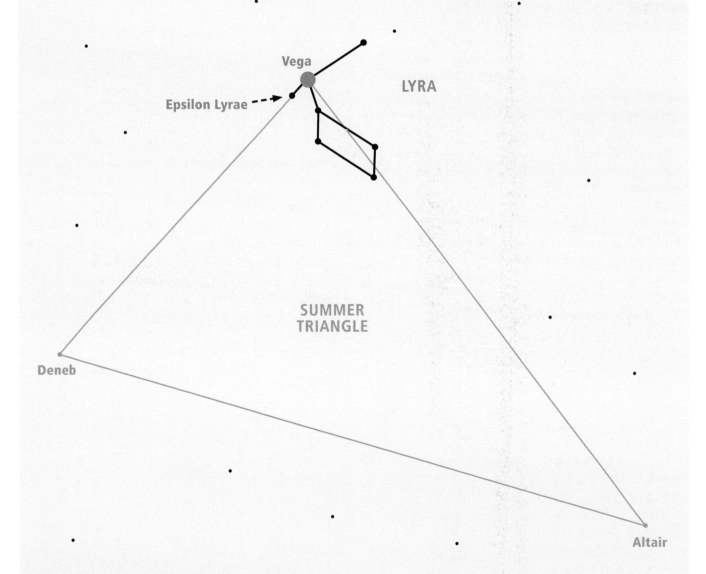

# 56

# LYRA

## *The Harp*

| CLASSIFICATION: **CONSTELLATION** | VISIBILITY: **MODERATE** |
| --- | --- |

Vega

Epsilon Lyrae

LYRA

SUMMER
TRIANGLE

Deneb

Altair

According to some ancient Greek legends Lyra was the harp of Orpheus, the best musician who ever walked the Earth. His music was so sweet and pure that even the trees bent over to listen. The rivers ceased flowing, wild beasts became tame, even mountains listened with pleasure when Orpheus played his magical music. One day Orpheus met a beautiful nymph named Eurydice. She too was mesmerized by Orpheus's music, and they fell in love and got married. Unfortunately, soon after the marriage Eurydice was bitten on the heel by a serpent and died. Orpheus was deeply saddened by her death and vowed to never play music again.

The gods, missing his sweet music as much as any mortal, came to Orpheus and allowed him safe passage to Hades to retrieve his lost love and bring her back to the land of the living. Through song, Orpheus convinced the god of the underworld to release Eurydice on the condition that he could not look at her until they emerged into the land of the living. But as they left Hades, Orpheus couldn't help himself and when he looked back for Eurydice he saw her being dragged back to the underworld never to be seen again. Orpheus's harp, the constellation Lyra, now remains in the sky as a reminder of true love, love lost, and why death is so hard to cheat.

Although small, this constellation holds many bright stars including the fifth brightest star in the sky, Vega. At the start of summer you will find Vega sparkling like a blue-white jewel at the top corner of the Summer Triangle. Four fainter stars form a neat, little parallelogram-shape that seems to hang off Vega by an invisible string. The parallelogram is the body of the harp and so "strings" are

definitely appropriate imagery for this star pattern.

### HOW TO FIND IT

**1** At the start of **SUMMER** look for Lyra's brightest star, Vega. From a darker sky the outline of the harp is more apparent. You may also see a sixth star with the naked eye close to Vega. That is called Epsilon Lyrae and is a famous quadruple star system. Astronomers also call Epsilon Lyrae the Double-Double, since you can see two tightly knit pairs of stars through a telescope. When you can make out the entire harp in the sky, the view of this tiny constellation is as sweet as Orpheus's music.

**2** In early **SUMMER** Lyra is the first constellation within the Summer Triangle to rise, and it marks the apex of the asterism. By late **SUMMER** and early **FALL**, Lyra is almost directly overhead after dark. And as the calendar moves toward **NOVEMBER** and **DECEMBER**, Lyra hangs in the western sky after dark. It may be known as a **SUMMER** constellation, but Lyra is still observable every evening well into **JANUARY**.

## 57 | VEGA

### The Star of Summer

CLASSIFICATION: **STAR** | VISIBILITY: **EASY**

Worshipped in ancient Egypt as early as 6000 B.C., and with its prominent place at the roof of the heavens, Vega is definitely the star of the summer skies.

Vega is a blue-white giant star that burns at well over 16,800 degrees Fahrenheit on its surface—much hotter than our 10,000-degree yellow Sun. Vega is one of our closer stars, only about 25 light-years from Earth. Our solar system is constantly moving in the direction of Vega at a rate of 12 miles per second. Even though the Sun, Earth, and all the planets are traveling through space at tremendous velocity, it will still take over 500 million years for our two stars to pass close to one another.

Fourteen thousand years ago Polaris was not the North Star—Vega was. Earth wobbles like a top, but it takes about 26,000 years for it to complete a single wobble. So over millennia our North Star changes depending on the pitch of the wobble. Today our North Pole points toward Polaris, but in another 12,000 years (around the year A.D. 14000) Vega will once again be the North Star.

#### HOW TO FIND IT

**1** Vega is the brightest star in not only the constellation Lyra but also the asterism of the Summer Triangle.

**2** It's a safe bet that on any given **SUMMER** evening that Vega, with the exception of possibly a planet, will be the brightest twinkling celestial body you see.

**3** You can first spy Vega during the month of **MAY** in the northeastern sky after dark. It forms the top corner of the Summer Triangle in **MAY**, **JUNE**, and **JULY**. Each month it will appear higher in the east until it is nearly straight overhead in **AUGUST** and **SEPTEMBER**.

**4** During the **FALL** months of **OCTOBER** and **NOVEMBER**, Vega will be lower in the western and northwestern sky until it leaves the evening sky altogether in **DECEMBER**.

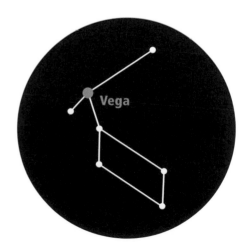

"Once I got into space, I was feeling very comfortable in the universe. I felt like I had a right to be anywhere in this universe, that I belonged here as much as any speck of stardust, any comet, any planet."

Mae Jemison, astronaut

# CYGNUS

## *The Swan*

| CLASSIFICATION: **CONSTELLATION** | VISIBILITY: **EASY** |

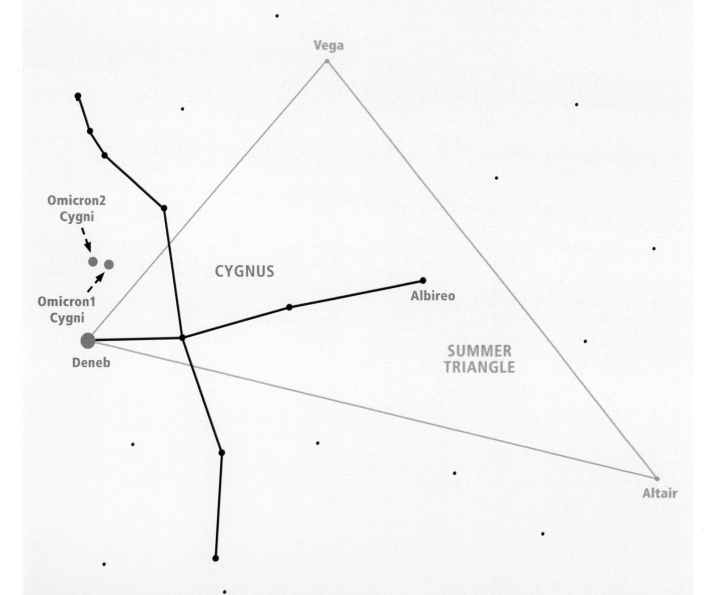

Vega

Omicron2
Cygni

CYGNUS

Albireo

Omicron1
Cygni

SUMMER
TRIANGLE

Deneb

Altair

One Greek legend describes Cygnus as the chief god himself, Zeus. One day Zeus fell in love with the queen of Sparta, a mortal woman named Leda. To woo the queen, Zeus disguised himself as a swan. Each day this strange yet compelling swan would appear at Leda's window. Eventually she thought, "My, there is something very attractive about that swan." Maybe it was the feathers or the way he threw thunderbolts, but as strange as it sounds, Leda fell in love with the swan. From this divine yet unholy union Leda became pregnant and, yes, laid an egg. When the egg cracked, out popped the Gemini twins, Pollux and Castor.

Cygnus, the Swan, is one of the easier constellations to picture in the sky. It does not take a lot of imagination to see a bird flying through the heavens in these stars. Start from the tail star Deneb and move inside the Summer Triangle to find fainter stars in a line that form the body and two windswept wings. A star named Albireo represents the swan's head at the end of a long neck.

Instead of a swan many modern stargazers have given Cygnus a nickname: the Northern Cross. The line between the stars Deneb and Albireo is the long part of the cross, while the line of three stars making the swan's body and wings form the shorter part of the cross. Although it is mostly known as a summer constellation, Cygnus's stars can be visible well into the evening skies of December. At that time of year it stands upright like a cross above the northwestern horizon as a symbol for the Christmas season.

The swan is located in a rich part of the Milky Way. In some early cultures the Milky Way was considered to be the river in the sky and the swan, a water bird, naturally flew along it. The Milky Way appears to split into two channels in Cygnus with a dark, apparently starless area in the middle. That darker rift in the Milky Way is called the Coalsack.

### HOW TO FIND IT

**1** To find Cygnus, look for its brightest star, Deneb, securing one corner of the Summer Triangle and indicating the position of Cygnus's cross-like shape. As **SUMMER** begins Deneb and Cygnus anchor the left side of the Summer Triangle when it rises in the northeastern sky.

**2** As **SUMMER** progresses, Cygnus will rise higher in the evening sky each night, and by **SEPTEMBER** and **OCTOBER** it will be almost straight overhead and stand at the top of the Summer Triangle. By **NOVEMBER** and **DECEMBER** it will get lower in the western sky, but it will remain visible through the month of **JANUARY** just after sunset in the northwest, on the right side of the Summer Triangle.

## 59 | DENEB

*The Tail of the Swan*

| CLASSIFICATION: **STAR** | VISIBILITY: **EASY** |

The name Deneb comes from the Arabic word for "tail." This star shows you the tail end of Cygnus, the Swan. Deneb shines with a blue-white light and scorches space with a surface temperature of about 14,800°F. Not only is it the farthest star in the Summer Triangle, Deneb is perhaps the farthest star you can see with the naked eye. Although sky surveys vary in their measurement of Deneb's distance from Earth, some astronomers think it could be more than 2,600 light-years away. That's the equivalent of about 15,000,000,000,000,000 miles! The fact that we can still see it as a bright star must mean that Deneb is humongous. By some estimates Deneb could be about 19 times more massive, over 200 times wider, and 200,000 times brighter than our Sun. If Deneb was hundreds of times closer to Earth—like at the equivalent distance in which Sirius resides—it would completely light up the nighttime sky.

Because of its great distance from us, Deneb appears to be the faintest star in the Summer Triangle and is also the farthest north of the three great brilliant stars of this asterism.

### HOW TO FIND IT

**1** While the brighter star Vega rises in the northeast in **MAY**, dimmer Deneb emerges from the north-northeastern horizon a few hours later and marks the left corner of the Summer Triangle. It will be about 24 degrees down and to the left of Vega.

**2** In **SEPTEMBER**, Deneb will be the top corner of the Summer Triangle. At that time of year it rides near the zenith and is so high overhead that you will need to lie down to observe it. At that point Vega will be 24 degrees below Deneb.

**3** By **DECEMBER** and **JANUARY** this "summer" star is still fairly high up in the western sky and will mark the upper-right corner of the Summer Triangle.

**4** Some eagle-eyed observers can still catch Deneb right before sunset through the end of **FEBRUARY**. It will be low above the north-northwestern horizon an hour after Vega has already set.

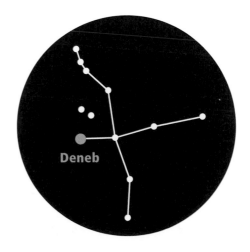

Deneb

# 60 | ALBIREO

*The Head of the Swan*

CLASSIFICATION: **STAR** | VISIBILITY: **MODERATE**

From about 430 light-years away Albireo looks like an ordinary white star, but through even a modest telescope you can discover that Albireo is in fact a beautifully colored double star (two suns that may orbit each other). The brighter of the two stars is orange with a dimmer blue star next to it. They make a stunning pair and are regarded by amateur astronomers as one of the best displays of contrasting star colors. It doesn't look like it, but the pair is separated by a vast distance. If they do revolve around each other, it may take up to 100,000 years for them to complete one orbit.

## HOW TO FIND IT

**1** To find Albireo, first find the stars that comprise the Summer Triangle, and then take a closer look at the faintest of the three, Deneb. If Deneb is the tail of Cygnus, then the swan's body, wings, and head lie inside the Summer Triangle. From Deneb travel 22 degrees inside the triangle along three fainter stars, and at the end you will find Albireo. It is also 15 degrees away from Vega and 20 degrees from Altair.

**2** Albireo lies at the western end of the constellation Cygnus and forms the swan's head. Some cultures referred to it as the "beak star" since it fits perfectly at the end of the swan's stretched-out neck.

**3** Along with Cygnus, Albireo rises in the northeastern sky in **JUNE** and **JULY**, travels nearly straight overhead in **AUGUST**, **SEPTEMBER**, and **OCTOBER**, and then goes lower in the northwestern sky just before **WINTER** begins.

**4** Some people picture this star pattern as a cross rather than a swan. The cross shape is definitely distinct once you find its place on the Summer Triangle. Connect the dots between Deneb and Albireo to make the longer side of the cross, and then connect other stars (Gienah, Sadr, and Delta Cygni) to form the shorter side. In **DECEMBER** you can see the Northern Cross standing upright in the northwestern sky—Deneb at the top, Albireo at the bottom.

# 61

# OMICRON1 CYGNI AND OMICRON2 CYGNI

## *Seeing Double*

CLASSIFICATION: **STARS** | VISIBILITY: **MODERATE**

Near Deneb you may be able to detect two stars that are just barely visible to the naked eye. They appear to be so close together that you might need to look twice. These stars are Omicron1 Cygni and Omicron2 Cygni (astronomers use the word *Cygni* to designate stars within the constellation Cygnus). They appear to be only 1 degree apart in the sky and are almost identical in brightness. The two stars may look close together but they are worlds apart. Omicron1 is about 880 light-years away and Omicron2 is a whopping 1,100 light-years from Earth. If every light-year is the equivalent distance of 5.8 trillion miles, that means these stars are 1.3 quadrillion miles apart.

### HOW TO FIND IT

**1** The trick to identifying all the stars in Cygnus is to start with Deneb. This tail star in the swan constellation is the dimmest of the three Summer Triangle stars and the farthest to the north.

**2** After you identify Deneb, look for the cross-like pattern of stars that form the body of the swan. Deneb is the tail, three stars in a line to Deneb's right form the wings.

**3** You can find Omicron1 and Omicron2 halfway between Deneb and Cygnus's left wing star, which is called Delta Cygni. Since they are much dimmer than Deneb, they will appear as two white points of light that are very close together.

**4** Omicron1 and Omicron2 Cygni look even better through a pair of binoculars. Scan the sky between Deneb and Delta Cygni and you can't miss them. They'll look a little like the eyes of an owl staring back at you through the darkness.

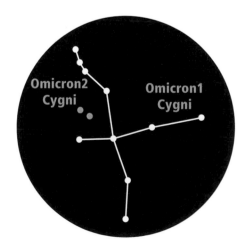

"A philosopher once asked, 'Are we human because we gaze at the stars, or do we gaze at them because we are human?' Pointless, really… 'Do the stars gaze back?' Now, that's a question."

Neil Gaiman, author

| # AQUILA

## *The Eagle*

| CLASSIFICATION: **CONSTELLATION** | VISIBILITY: **MODERATE** |
|---|---|

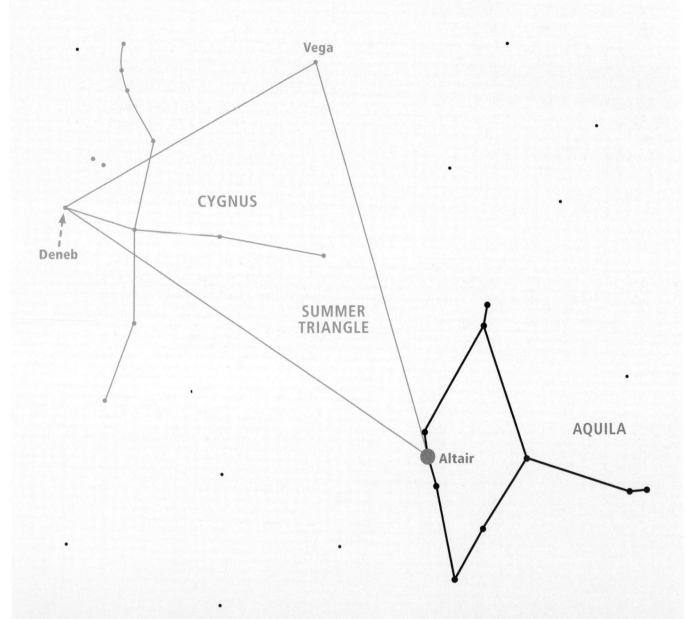

Many Greek and Roman myths indicated that Aquila, the Eagle, carried the thunderbolts that the god Zeus or Jupiter frequently hurled at troublesome humans. In fact, since eagles seemed to be the fastest creatures, the Greeks associated them with lightning strikes.

The Greeks created several myths around this bird of prey: Some stories gave the eagle very mundane tasks to do while other tales were downright terrifying. In one legend Zeus was sitting on Mount Olympus and developed a mighty thirst. Instead of getting a drink himself, he sent his eagle to fetch a mortal from below to do the job. The eagle raced down to the city of Troy and snatched up a young prince named Ganymede and brought him back to Mount Olympus. Ganymede, far removed from being a prince, then became a servant to Zeus as his cupbearer.

In another tale, the eagle became an instrument of torture. The god Prometheus developed a soft spot for humans and how they suffered on Earth. He decided to introduce fire to humanity and help them get through bad weather. Zeus, fearing the humans would now become too powerful, punished Prometheus for this act by chaining him to a rock. When the Sun rose each morning, Zeus sent his eagle to sit on Prometheus's lap and eat out his liver. At the end of the day, the eagle would fly home and Prometheus's liver would magically grow back. As the Sun rose, the eagle returned, and this punishment went on day after day for eternity.

## HOW TO FIND IT

**1** Beyond the dazzling star Altair, Aquila has very few bright stars, and its outline is tough to discern. It may be drawn as a diamond or a chevron shape, depending on the artist.

**2** The easiest way to find Aquila is to locate Altair in its far-flung position in the Summer Triangle. While Vega and Deneb are only 24 degrees apart, Altair is the farthest star from the other two (34 degrees from Vega and 38 degrees from Deneb). Aquila is also the southernmost member of the Summer Triangle. This means it is the last to emerge above the treetops, does not travel as high up in the sky as Vega and Deneb, and is the first to set.

**3** At the beginning of **SUMMER**, look for the eagle rising in the east just after dark. Deneb will be over to the left while Vega is perched atop the triangle. By **AUGUST** and **SEPTEMBER**, Aquila will fly about halfway up in the southern sky. During **NOVEMBER** and **DECEMBER** the eagle will be lower in the west.

**4** At the start of **WINTER** the entire Summer Triangle will be visible just after sunset. It will still exhibit the same shape, but the perspective will be different. Altair will be on the left, Vega will be on the right, and Deneb will be at the top of the triangle. After Aquila leaves the sky, Lyra and Cygnus will soon follow.

## 63 | ALTAIR

### *The Eagle Eye*

| CLASSIFICATION: **STAR** | VISIBILITY: **EASY** |

Vega might be the brightest star in the Summer Triangle, but it is not the closest. That honor goes to Altair, which is only 17 light-years from Earth. It's so close (relatively speaking) that we can actually study Altair in better detail than most stars in the galaxy. Using the best telescopes in the world, astronomers are able to detect Altair's rapid rotation. While the Sun spins once every 25 to 26 days, Altair completes one turn every 9 hours. This rapid rotation has flattened it at the poles and sent more mass to its equator. So it looks more like a blue-white piece of Skittles candy than a spherical ball of gases.

An old Korean legend links Altair to the star Vega in the constellation Lyra. In this story two star-crossed lovers, a cowherd and a weaver, were banished to the heavens. Their love was so strong that they fell into the sky hand in hand. It looked as if they would finally be together until a flock of magpies flew in between the two. Their hands separated and they glided away from each other. When they stuck to the heavens, the weaver was on one side of the great river in the sky (the Milky Way) and the cowherd fell on the opposite shore. The cowherd turned into the star Altair and the weaver turned into Vega. The couple could meet only once a year on the seventh day of the seventh month. Only then could they cross the river with the help of friendly magpies.

Magpies in Korea flocked like crazy in July and, to atone for their part in this tragic tale, were said to fly up to the stars to construct a bridge across the river for the couple to cross. Legends describe that when the meeting occurs on July 7, Altair and Vega shine in five colors to symbolize their happiness. If it rains on July 7, so the legend goes, it is a sad night, for the couple fails to meet at all.

### HOW TO FIND IT

1. Altair is the eagle eye of the constellation Aquila, the Eagle, and is the second brightest star in the Summer Triangle.

2. Once you identify the Summer Triangle, Altair is the star farthest from the other two and appears farthest to the south. It shines with a stark white light.

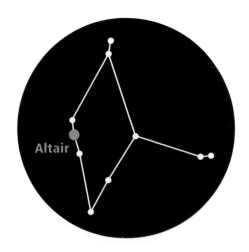

"To explain all nature is too difficult a task for any one man or even for any one age. 'Tis much better to do a little with certainty, and leave the rest for others that come after you, than to explain all things by conjecture without making sure of anything."

Isaac Newton, mathematician and physicist

# DELPHINUS

## *The Dolphin*

| CLASSIFICATION: **CONSTELLATION** | VISIBILITY: **MODERATE** |

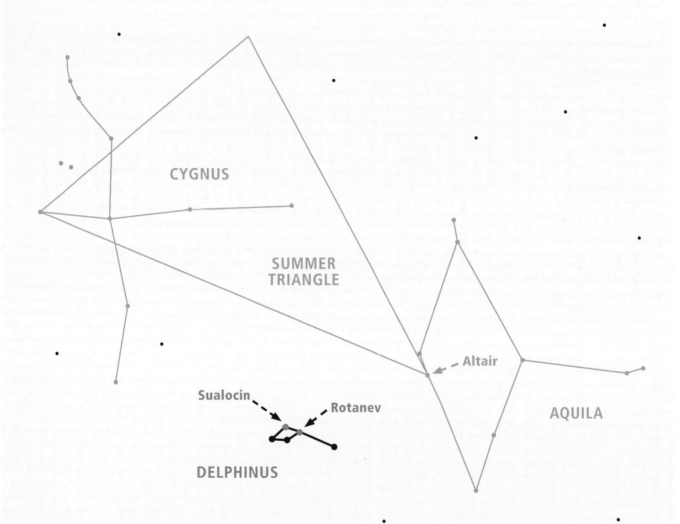

CYGNUS

SUMMER
TRIANGLE

Altair

Sualocin

Rotanev

AQUILA

DELPHINUS

Delphinus, the Dolphin, is one of the smallest constellations in the summer sky, and it lies just outside the Summer Triangle. Once you find it, you may even be persuaded to agree it looks a little like a dolphin arching its back and jumping above the cosmic waves. According to Greek myths, Delphinus was said to be the messenger of Poseidon, the god of the sea. Delphinus won great acclaim for saving the life of Arion (Poseidon's son) when his ship was attacked at sea. Ancient mariners often attributed the dolphin with great wisdom and also a love for humans. Dolphins were thought to be sailors' best friends and legends swirled through the Mediterranean (both real and fictional) about how they had rescued humans and gotten shipwrecked fools out of trouble.

Delphinus also helped Poseidon get a date with the beautiful Nereid named Amphitrite by delivering her a message when she was in the Atlas Mountains. I'm not sure how this dolphin-gram reached her, since he couldn't exactly swim it up the mountain, but that would be an interesting sight!

The great river in the sky, the Milky Way, is not far from Delphinus. The cloudy band of stars flows just above the dolphin's head and spans the area between the constellations Cygnus and Aquila.

## HOW TO FIND IT

**1** Delphinus consists of third and fourth magnitude stars, so it will be a challenge to find this little constellation if you live in a city. Even if you live out in the country, this star pattern won't jump out at you. You will see it out of the corner of your eye as a little faint glow, but upon closer examination you should be able to make out the distinctive arch of stars that covers only about 5 degrees of the **SUMMER** sky.

**2** To find Delphinus, look along the longest side of the Summer Triangle (between the stars Altair and Deneb). Just outside the triangle and closer to Altair search for a small diamond shape of four stars with one or more stars off to the right. The stars in the diamond mark the body, and the extra star to the right is the tail.

**3** It's easier to spot the Dolphin when it is higher in the sky. That doesn't happen until evenings in **JULY** when it rises higher above the eastern horizon after dark. In **AUGUST**, **SEPTEMBER**, and **OCTOBER** you can find Delphinus high in the southern sky. By **OCTOBER** and **NOVEMBER** it hangs out in the west and is pretty much gone from the evening sky from **DECEMBER** until **JUNE**.

# SUALOCIN AND ROTANEV

## *Backward and Forward*

| CLASSIFICATION: **STARS** | VISIBILITY: **MODERATE** |

You have read about stars named by the Greeks, Romans, and Arabs, but the two brightest stars in Delphinus—Sualocin and Rotanev—are a little more modern and are actually named after one particular person, a man named Niccolo Cacciatore.

Cacciatore was an assistant to the Sicilian astronomer Giuseppe Piazzi, the discoverer of the first and largest asteroid: Ceres. In 1814 Cacciatore helped Piazzi prepare a new star catalog. When other astronomers read the work, they found two stars in Delphinus had acquired new and very unique names. Only decades later did anyone figure out that the names, Sualocin and Rotanev, were the reverse spelling of Nicolaus Venator, the Latinized version of the name Niccolo Cacciatore. Niccolo snuck his own name into the stars of Delphinus! And the names have stuck ever since and show a most creative way to become immortalized in the stars.

Sualocin and Rotanev make up the arching back of the dolphin constellation. Sualocin, on the top of the diamond shape, is blue-white in color and about 241 light-years from Earth. Rotanev, on the right side of the dolphin, is a white star about 101 light-years away.

### HOW TO FIND IT

**1** To find Delphinus's uniquely named stars, scan along the longest side of the Summer Triangle (between the stars Altair and Deneb). Just outside the boundaries of the triangle and closer to Altair you may find a small diamond shape that forms the body of the Dolphin.

**2** Sualocin and Rotanev are the brightest two stars in the diamond and should shine with a similar white glow. Sualocin is the top star in the diamond with Rotanev just 1.5 degrees to the right.

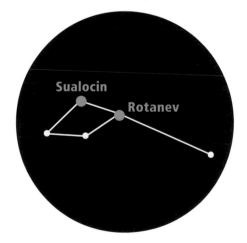

"The important thing is not to stop questioning. Curiosity has its own reason for existence. One cannot help but be in awe when he contemplates the mysteries of eternity, of life, of the marvelous structure of reality. It is enough if one tries merely to comprehend a little of this mystery each day."

Albert Einstein, theoretical physicist

# 66

# OPHIUCHUS

*The Serpent Bearer*

| CLASSIFICATION: **CONSTELLATION** | VISIBILITY: **MODERATE** |

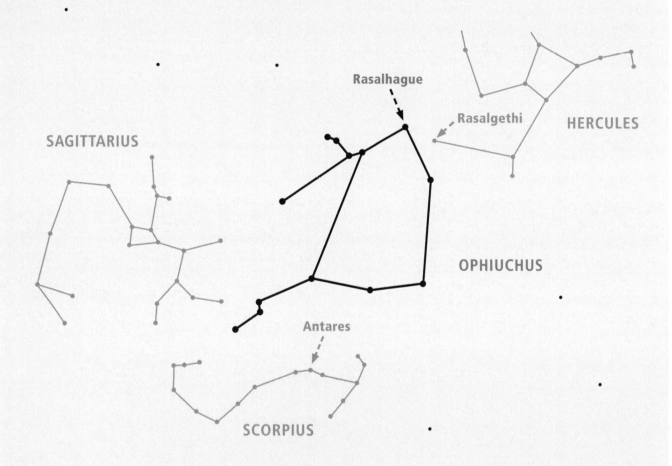

Many legends say that this constellation shows a man holding a huge serpent. The Greeks believed this guy was named Asklepios, instead of Ophiuchus. He was the god and inventor of medicine and was so gifted that he could even restore the dead to life.

Asklepios was the son of the god Apollo and a mortal princess named Coronis. He was mentored in the field of medicine by the centaur Chiron (some say Chiron is embodied in the nearby constellation Sagittarius). Once when Asklepios was making a house call to a sick patient, a serpent slithered into the room and coiled around a staff. Asklepios, slightly scared of snakes, quickly killed it. A few minutes later a second serpent crawled under the door and entered the room carrying an odd herb in its fanged mouth. The second serpent went over to the dead one and applied the herb, restoring the dead serpent to life. From that moment on Asklepios always carried a staff with a serpent wrapped around it. It became the symbol for the medical arts still in use today.

Asklepios brought many people back from the dead, including Orion. As described in the Orion myth, the gods sent a scorpion to humble Orion—by killing him. In one alternate ending of the myth Asklepios was called to the scene of the crime to work his doctorly deeds. Not only did Asklepios raise Orion to life but he even dispatched the scorpion by squishing it under his sandaled foot.

The constellation Ophiuchus looks like a long stretched-out pentagon of stars, but none of them are terribly bright. The star named Rasalhague (meaning "head of the serpent charmer") is the brightest star in Ophiuchus and marks the top of the pentagon. Rasalhague is a white star lying about 49 light-years from Earth.

Just to the right of Rasalhague is a star in the constellation Hercules called Rasalgethi, or "head of the kneeler." So, very close together, we have two men bumping heads in the heavens! But Ophiuchus is lucky. He gets to stand right-side up, while Hercules flies through the sky upside down.

## HOW TO FIND IT

**1** To find the elongated pentagon shape of Ophiuchus, search about 25 degrees above the constellation Scorpius and about 25 degrees below Hercules. You can also locate Ophiuchus by finding its brightest star, Rasalhague, which marks the apex of the pentagon shape. When Ophiuchus stands upright halfway up in the southern sky during the **SUMMER** months, Rasalhague is bracketed by two very bright stars. Look for Antares in Scorpius, the Scorpion, lower in the south, and Vega from the Summer Triangle high above it. If you draw a line from Antares up to Vega (a jump of about 70 degrees), you'll find Rasalhague on that line about 40 degrees above Antares.

**2** The best months to look for Ophiuchus are **JULY**, **AUGUST**, and **SEPTEMBER** when he stands about halfway up in the southern sky. Remember that he is stepping on the scorpion, so you should notice the more prominent stars in Scorpius below his feet in the southern sky.

# 67

# SCORPIUS

## *The Scorpion*

| CLASSIFICATION: **CONSTELLATION** | VISIBILITY: **EASY** |

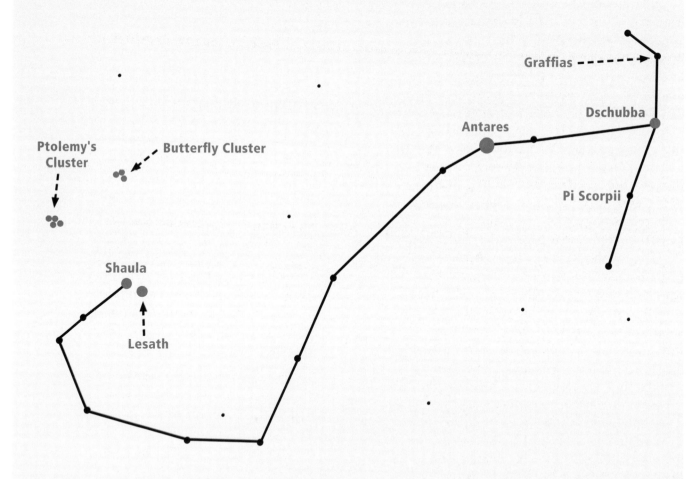

Scorpius is a very distinct constellation that lies low in the southern sky on summer evenings. From most locations in the mid-northern latitudes it appears to creep across the heavens just above the treetops. The brightest stars in Scorpius form the shape of a fishhook or the letter J. This is the body and tail of the scorpion, and out of all the constellations in the sky this group of stars may look the most like its namesake.

The fishhook and brightest star, Antares, are definitely Scorpius's most notable features, since the head and claws of the scorpion can be very obscure. The end of the scorpion's tail has two stars, Shaula and Lesath, that are uniquely close together. From most places in the United States these tail stars appear so low in the southern sky that you will have a very short window to see them above the treetops.

Mimicking Orion's Belt, Scorpius has three stars in a similar line at its head. The stars are not quite as bright and not quite as aligned as Orion's Belt, but they are definitely noticeable. The star at the top of the line is named Graffias, the star in the middle is Dschubba, and the one at the bottom is Pi Scorpii. The scorpion's claws that reach out to the right are very difficult to imagine unless you're viewing a dark sky. Indeed, on many star charts the scorpion's claws are actually part of the nearby constellation Libra, the Scales. The claws' stars that can be found in Libra bear the memorable names Zubenelgenubi and Zubeneschamali. Although astrology refers to this zodiac sign as Scorpio, you should use its more astronomical moniker: Scorpius. This is the killer of Orion, the dreaded scorpion that stung the hunter on his heel and sent him to the afterlife among the stars.

Legend has it that after Orion died the gods offered him a place in the sky to live on as a constellation. Orion made one simple request: He never wanted to see that scorpion again. The gods honored this wish and placed him on the opposite side of the firmament from the scorpion. The two constellations are never in the sky at the same time. Orion rules the winter sky, and when he rises in the east Scorpius has already set in the southwest. During the summer months Scorpius emerges into the sky only after Orion has left the scene and has drifted below the western horizon.

## HOW TO FIND IT

**1** The stars in Scorpius make a very distinct fishhook-shaped pattern in the nighttime sky. You don't need any pointer stars to find it because if you look to the south on a **SUMMER** evening, you can't miss it.

**2** The scorpion constellation lies so far to the south that you only have a small window of time to spy it before it goes below the horizon. In fact, since it never gets very high in the sky it always looks like it is crawling above the treetops from southeast to southwest, which only adds to its mystique.

**3** Scorpius starts to become visible in the evenings in late **JUNE** when it rises in the southeastern sky after dark. In **JULY** and **AUGUST** the scorpion reaches its highest evening position in the southern sky but is still only about 20 degrees above the horizon. And by **SEPTEMBER** it is already difficult to see Scorpius very low in the southwest just after sunset.

## 68 | ANTARES

*The Heart of the Scorpion*

| CLASSIFICATION: **STAR** | VISIBILITY: **EASY** |

Antares is one of the reddest stars visible to the naked eye. Because it often hovers just above the horizon, it twinkles a lot. It also just happens to mark the heart of the constellation Scorpius, and all that twinkling can create an impression of a wildly beating heart in the middle of the scorpion's body.

Antares is a red supergiant star and one of the largest known stars in the galaxy. If Antares was our sun, it would engulf the entire orbit of Mars. Earth would be well inside it. But luckily Antares lies about 620 light-years away.

The name Antares comes from the ancient Greek word meaning "rival of Mars," because of its similar color to the Red Planet. The Chinese called it the Fire Star for the same reason. Occasionally the planet Mars passes near Antares and you can see them side by side. Although they may seem to be the same color, Mars, as a planet, does not twinkle nearly as much as Antares.

As another link to the constellation Orion, Antares has many similarities to Betelgeuse, the star marking the hunter's armpit. Astronomers are watching and waiting for Antares to suddenly flare up into a massive supernova. Which star will blow up first? Antares or Betelgeuse? The answer to that is anyone's guess. But when Antares goes supernova, the light we see from this star may even outshine a Full Moon.

### HOW TO FIND IT

**1** To find Antares, look for the brightest star in Scorpius marking the creature's red, beating heart.

**2** Antares is visible in the evening sky between **JUNE** and **SEPTEMBER**. When you are watching a fireworks show at night on the Fourth of **JULY**, and the skies are cloud free, you may take special note of this red star twinkling like crazy amid the rockets' red glare.

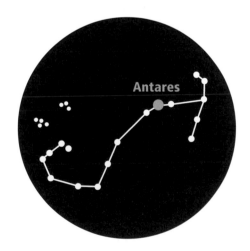

## 69 | DSCHUBBA

### The Forehead

| CLASSIFICATION: **STAR** | VISIBILITY: **MODERATE** |

Dschubba is an Arab-named star that means "forehead," which is where Dschubba sits—to the right of the heart star Antares, and toward the head of Scorpius, the Scorpion.

It is a blue-white star more than 400 light-years from Earth. In the summer of 2000, the ever-steady star Dschubba suddenly and unexpectedly brightened. This led astronomers to take a closer look, and what they found was that Dschubba was not one star but a system of up to four stars. The interactions of the four stars could have sparked the flare-up. And now the star is significantly brighter than it once was just a few decades ago.

Dschubba also has a dense ring of material surrounding it. Astronomers are not sure where this debris came from but hypothesize that this star spun so rapidly that parts of it spiraled out into space and now circle Dschubba like a swarm of planets.

**HOW TO FIND IT**

1  First locate Scorpius. Look for the fish-hook formation of stars low in the southern sky. The bright red star Antares marks the scorpion's heart while the fainter tail and stinger stars trail off to the left.

2  About 7 degrees to the right of Antares you'll find three semi-bright stars that are almost in a row. They look a little like a crooked version of Orion's Belt. Those three stars are supposed to be the head of Scorpius. The star on the top is called Graffias and the one on the bottom (and the dimmest of the three) is Pi Scorpii. The middle star and brightest of the three is Dschubba.

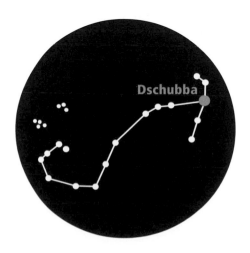

# SHAULA AND LESATH

*The Sting and Bite*

| CLASSIFICATION: **STARS** | VISIBILITY: **MODERATE** |

The two stars at the tip of the tail of Scorpius, the Scorpion, are bright and appear extremely close together. The brighter one is a big blue star named Shaula, which means "the sting" in Arabic. The fainter blue-white one is Lesath. The origin of this name is shrouded in mystery. It could be a bad translation from the Arabic word meaning "bite of a poisonous animal." Or it could be from a rough Greek translation from a word meaning "foggy," since it lies near two open star clusters named M6 and M7 that give off a foggy glow in the sky.

Although they appear close together in the sky, Shaula and Lesath are separated by trillions of miles. Shaula is about 570 light-years away while Lesath is just a little farther at 580 light-years from Earth. If a difference of 10 light-years sounds close to you, it isn't. One light-year is the distance light travels in a year—the equivalent of 5.8 trillion miles. So these two stars are not exactly neighbors in space.

## HOW TO FIND IT

**1** To locate Shaula and Lesath, first find Scorpius and its fishhook pattern of stars low in the southern sky. The brightest star in the constellation, Antares, marks the scorpion's heart.

**2** After you find that, follow Scorpius's body down to the left, away from Antares. The stars will then start to curve upward like the end of a fishhook. This is the scorpion's tail and stinger that killed mighty Orion. The brighter star here is Shaula with dimmer Lesath right next to it. The two stars appear to be only ½ of a degree apart. That means that if you extend your arm all the way, you can cover both stars with the tip of your pinkie.

**3** Shaula and Lesath only appear low in the southern sky after dark. So the biggest obstacle to seeing them is trees. For the best results try viewing them from a location with a clear view to the southern horizon. The best months to search for them are **JULY** and **AUGUST** when they get 10–20 degrees above the southern horizon.

# BUTTERFLY AND PTOLEMY'S CLUSTERS

## Two Little Clouds of Stars

| CLASSIFICATION: **STAR CLUSTERS** | VISIBILITY: **DIFFICULT** |
|---|---|

The Butterfly Cluster and Ptolemy's Cluster have been known since antiquity, and they are just barely visible to the naked eye. They are also named M6 and M7, respectively.

Ptolemy's Cluster (M7) is the brighter and closer of the two clusters in Scorpius. The combined light of the 80 stars in this grouping are nearly equivalent to a third magnitude star, and they all lie almost 1,000 light-years from Earth. The cluster is named after the ancient Greek astronomer Ptolemy, who wrote some of the most important and influential astronomy texts in history. In A.D. 130 Ptolemy described this group of stars as the "nebula following the sting of Scorpius," since it seems so near to the scorpion's stinger stars, Shaula and Lesath.

Just above Ptolemy's Cluster, you may find the smaller and fainter Butterfly Cluster (M6). The stars in this group are estimated to be around 1,600 light-years from Earth and are still visible, although faintly, to a well-trained naked-eye observer. The entire cluster shines at around fourth magnitude, so you will need a dark sky to see it distinctly.

### HOW TO FIND IT

**1** M6 and M7 are best seen during **SUMMER** evenings and hang out low in the southern sky. First find the constellation Scorpius and its distinctive fishhook shape of stars. Follow the body of the scorpion down and to the left until you reach the two stinger stars that seem incredibly close together, Shaula and Lesath. Ptolemy's Cluster (M7) is just to the left of these stars. The Butterfly Cluster (M6) may be a tougher challenge to spot, but at best you may just be able to make it out as a little cloud just above M7.

**2** The two clusters look much more dramatic through a pair of binoculars or a small telescope, and by using them you can resolve individual stars quite clearly. Hunt around to the left of the scorpion's stinger and you should see them like a small swarm of fireflies in the **SUMMER** night. Does the shape of M6's stars vaguely look like a butterfly? You be the judge.

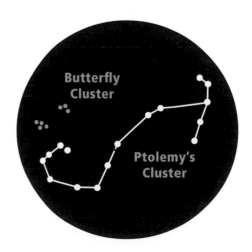

# LIBRA
## *The Scales*

| CLASSIFICATION: **CONSTELLATION** | VISIBILITY: **MODERATE** |

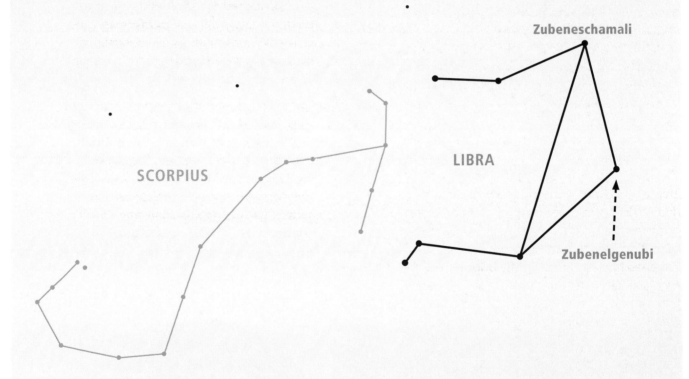

Zubeneschamali

LIBRA

SCORPIUS

Zubenelgenubi

Libra was the last zodiac constellation added to the twelve and the only one that was not a living creature. This diamond-shaped star pattern is supposed to represent the scales of justice that maintains the balance of law and order. The constellation came into vogue in ancient Rome when the ruling class had supreme respect for maintaining the rule of law.

The symbol for justice in the United States is a woman holding a set of scales. This may have come from a combination of Libra and the nearby constellation Virgo. In Greek mythology Astraea was the goddess of justice and sometimes stood in for Virgo among the stars. But according to classical illustrations of Virgo, the scales of justice can be found at her feet and not in her hand.

Scorpius is also near Libra, and the claws of the scorpion can be imagined pinching at the scales. Thousands of years ago the stars of Libra were often incorporated into the outline of the scorpion constellation, and Libra's brightest stars, Zubeneschamali and Zubenelgenubi, reflect this ancient connection. Zubeneschamali is the slightly brighter of the pair and is located higher in the sky. Its name means "the northern claw." Zubenelgenubi, standing below its fellow Libran star, means "the southern claw." In fact, the entire constellation of Libra used to mark the claws of a much larger scorpion. Only later did astronomers break Libra free from the scorpion's grasp and create a balance of power in the heavens.

## HOW TO FIND IT

**1** Look for this small constellation 10–15 degrees to the right of Scorpius. Libra's four main stars form a diamond shape, with Zubeneschamali at the top and Zubenelgenubi standing on the right corner of the diamond. Zubeneschamali and Zubenelgenubi are significantly brighter than the other two stars in the diamond shape, Gamma Librae and Sigma Librae, so you may not see the entire shape of Libra right away. But if you fix your gaze to the right of the scorpion you may still picture the stars of Libra as the scorpion's claws.

**2** Libra will rise before the scorpion and start to be visible in the southeastern sky in MAY. By JUNE and JULY it is about halfway up in the southern sky. And in AUGUST and SEPTEMBER Libra weighs the stars in the southwestern sky. The constellation is not visible in the evening skies between SEPTEMBER and MAY.

# 73

# SAGITTARIUS

*The Archer*

| CLASSIFICATION: **CONSTELLATION** | VISIBILITY: **MODERATE** |

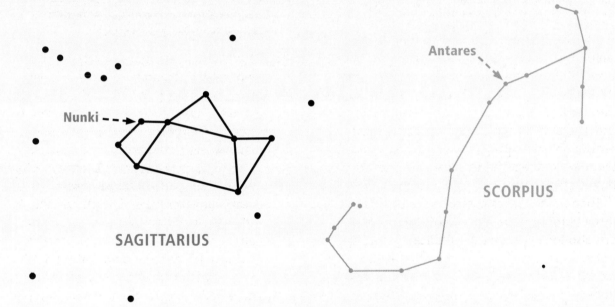

Nunki

SAGITTARIUS

Antares

SCORPIUS

The constellation Sagittarius represents Chiron, the Centaur, who is half man, half horse. Unlike most mythological centaurs that were wild, rude, and crude, Chiron was gentle, kind, and intellectual. His thirst for knowledge led him to become a great teacher of just about everyone in ancient Greece, including Hercules.

The shape of Sagittarius's stars resembles a teapot more than a creature that is half man, half horse. Like the fishhook pattern of stars that comprise Scorpius, the Scorpion, the teapot is another identifiable asterism—stars that are part of a greater constellation.

The star pattern of Sagittarius can be broken up into two smaller sections: the Bow and Arrow and the Milk Dipper. Look to the spout of the teapot pointing to the right to find Sagittarius's bow and arrow. Three stars curve to form the bow and one sticks out to form the arrow. Look at what he's aiming that arrow at! It is pointed directly at the Scorpion's heart, the red star Antares.

The four stars that comprise the handle of the teapot on the left side of the star pattern also look like a fainter replica of the Big Dipper's spoon. Some astronomers granted this formation the additional name "the Milk Dipper," because it lies in a thick patch of the Milky Way. When you are looking at Sagittarius, you are peering toward the center of our galaxy. Some imaginative stargazers picture the Milky Way as steam coming out of the spout of the teapot.

Sagittarius does not have any super bright stars, but try to locate Nunki, the brightest star in the Milk Dipper and the second brightest star in the constellation. It was named by the ancient Sumerians about 5,000 years ago, but today we have no idea what Nunki means. Maybe after observing this blue-white mystery star you can invent your own translation.

## HOW TO FIND IT

**1** If you can find Scorpius, you can find Sagittarius. Look for him prancing low across the southern sky, 15 degrees to the left of Scorpius's stinger.

**2** Sagittarius trails behind Scorpius in its nightly motion across the sky. The centaur rises a little later than the scorpion, and you can get a glimpse of the teapot-shape of stars later in JULY. Sagittarius is most easily spotted in AUGUST and SEPTEMBER where it reaches its highest point above the southern horizon after dark. However, even then the centaur is still very low in the sky—15–30 degrees above the treetops. By OCTOBER the point of his arrow is aimed at the setting Sun in the southwest, and the centaur quickly disappears from the evening sky until the following SUMMER.

# The Fall Sky

As the month of September arrives we find the stars and constellations of summer shifted ever westward in the sky. The Summer Triangle that you have been watching every summer evening is now tipped over on its side in the west. The scorpion has skittered to the southwest, and it is perched above the treetops and is about to set for the season. When you look east after dark a new group of constellations has arrived on the scene and has crept above the distant horizon. The stars of fall are not as bright as those in summer. Apart from the dazzling first magnitude star named Fomalhaut in the constellation Piscis Australis, the fall constellations are primarily constructed with second and third magnitude stars—whose muted brilliance can be a challenge to see from a city.

You cannot find your way around the fall sky without talking about the great fall sky saga. Every fall evening ancient Greeks looked up and saw a king, a queen, a prince, a princess, two monsters, and a flying horse, and they created an epic tale about these constellations. You have already read about the king and queen (the constellations Cepheus and Cassiopeia). They predominantly dwell in the northern sky and are visible throughout the year. But during the fall they are joined by their daughter Andromeda, her hero (the prince Perseus, who is holding the severed head of Medusa), the winged horse Pegasus, and the vicious sea monster named Cetus. These six constellations are spread over an area of the sky 100 degrees by 100 degrees. That means they cover fully one-quarter of the entire fall sky. If you can learn to navigate through these six constellations, then you will have an excellent foundation for identifying the major stars of the season.

In addition to the constellations in the fall sky saga, we'll also look at two smaller constellations of fall: Piscis Australis, the Southern Fish, and Aries, the Ram.

183

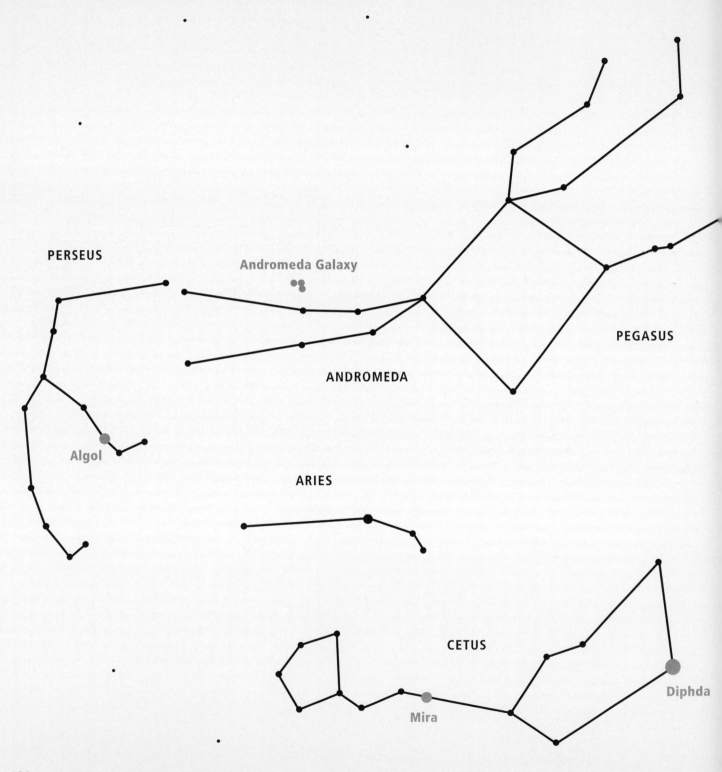

PERSEUS

Andromeda Galaxy

PEGASUS

ANDROMEDA

Algol

ARIES

CETUS

Diphda

Mira

# Fall Sky Constellations

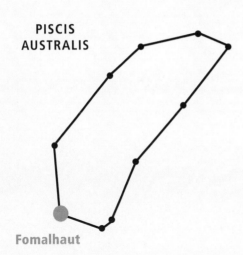

PISCIS
AUSTRALIS

Fomalhaut

# 74

# ANDROMEDA

## *The Princess*

| CLASSIFICATION: **CONSTELLATION** | VISIBILITY: **MODERATE** |

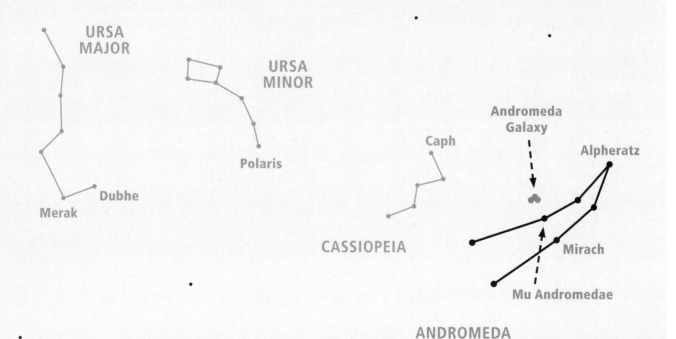

In Greek mythology Queen Cassiopeia bragged about her beauty all the time, but one day she offended the god of the sea, Poseidon. In this version of the story she said that her daughter, Andromeda, was more beautiful than all the mermaids in the ocean. When Poseidon heard about this he was outraged and stormed to the castle to confront the king and queen. "Being a fair and angry god," Poseidon said, "I will give you two choices for your punishment: I will send down a tremendous tidal wave on your land killing everyone and everything or I will take your one and only daughter, Andromeda, chain her to the big rock in the sea, and let my sea monster eat her."

Cepheus and Cassiopeia didn't have much choice. Either way their daughter was doomed. If the king selected wisely, he could at least save his kingdom from total destruction. Reluctantly, the queen's men rowed Andromeda out to sea, chained her to the sacrificial rock in the ocean, and waited. Poseidon's sea monster then emerged from the briny deep and was about to eat Andromeda. Who arrives just in time to save her? It's Perseus, our hero! Perseus aimed the head of Medusa at the sea monster and the monster immediately turned to stone. Andromeda was saved!

Andromeda's stars form a skinny letter A shape—A for *Andromeda*. The top of the A, or her head, is the semi-bright star called Alpheratz. Andromeda's leg stars point down to the northeastern horizon in early fall, but you will notice that the stars on the left side of her body are significantly brighter than the stars on her right side.

Alpheratz is a large blue-white star that lies about 97 light-years from Earth. The name comes from Arabic, meaning "the navel of the mare." And this is where it may get complicated. You will notice that not only does Alpheratz mark the princess's head but it is also one corner of the Great Square, the stars marking the body of Pegasus, the flying horse (see illustration in the Pegasus: The Flying Horse entry). So Andromeda's head doubles as Pegasus's rear. It is not a glamorous place for a princess, but that means Alpheratz will help you identify two constellations in the fall sky.

## HOW TO FIND IT

**1** To find Andromeda, you need to find the star Alpheratz. First locate the Big Dipper then connect the dots between the dipper's stars Merak and Dubhe. Continue on to Polaris, and then to Cassiopeia's star Caph. Keep going, make one more hop, and you will come to Alpheratz. Each hop is nearly equidistant (about 30 degrees in the sky), and each star along the way is nearly equal in brightness.

**2** In early **FALL** Andromeda seems to lie on her left side in the east-northeastern sky with her legs pointed downward. Later in the **FALL** Andromeda will appear almost straight overhead after sunset. She is visible well into the **WINTER** months as well. By **DECEMBER**, **JANUARY**, and **FEBRUARY** you can find her in the western sky, setting headfirst toward the horizon.

# 75

# ANDROMEDA GALAXY

CLASSIFICATION: **GALAXY** | VISIBILITY: **DIFFICULT**

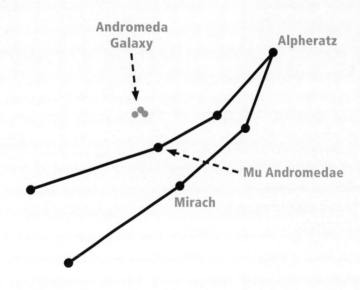

Andromeda
Galaxy

Alpheratz

Mu Andromedae

Mirach

ANDROMEDA

We now come to the farthest thing you can see with the naked eye: the Andromeda Galaxy. Also known as M31, the galaxy is composed of approximately 1 trillion stars that lie about 2.5 million light-years away. Imagine the brightness of our Sun multiplied 1 trillion times. Even though it emits that much light into the universe, you still would have trouble observing the Andromeda Galaxy from your backyard. That's how far away it is.

Stargazers throughout the Northern Hemisphere have noticed the Andromeda Galaxy for millennia without fully knowing what it was. In A.D. 964 Persian astronomer Abd al-Rahman al-Sufi observed the Andromeda Galaxy and called it a "little cloud" in his Book of Fixed Stars. This is probably the best description of what you can see of M31 on a good night.

Of course you can more easily spot M31 with a pair of binoculars. But even if you observe the Andromeda Galaxy through a telescope, you'll still see a fuzzy, cigar-shaped blur with a brighter central core. The view might not impress you at first, but remember that you are seeing 1 trillion stars shining at you from 14.5 quintillion miles away.

## HOW TO FIND IT

1. To find the Andromeda Galaxy, first locate the Big Dipper. Then connect the dots between the dipper's stars Merak and Dubhe.

2. Continue on to Polaris, and then to Cassiopeia's star Caph. Keep going and make one more hop and you will come to Andromeda's head star, Alpheratz.

3. Then follow along her body until you reach her hip stars, the brighter one named Mirach and the dimmer one called Mu Andromedae.

4. Connect the dots from Mirach to Mu Andromedae, and then continue that line for another 3–4 degrees. You will run right into the Andromeda Galaxy. The distance between Mirach and Mu, and Mu and M31 are about equal, so the hip stars will make you hip to the galaxy's location!

# 76

# PERSEUS

*The Hero*

| CLASSIFICATION: **CONSTELLATION** | VISIBILITY: **MODERATE** |
|---|---|

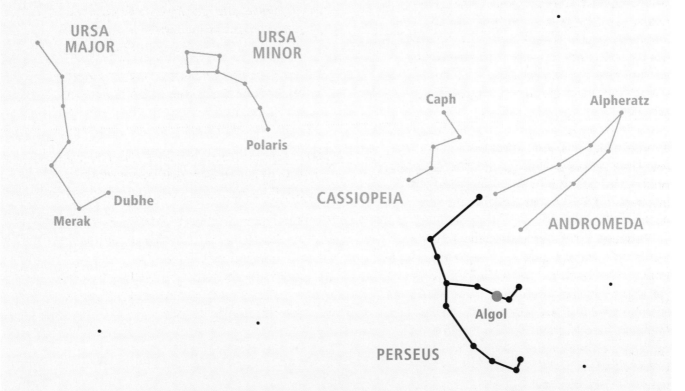

Perseus was the son of the Greek god Zeus and a mortal woman named Danaë. Favored by the gods, Perseus received some valuable gifts. From Hephaestus, the black-smith god, he got a sword that would cut through anything; from Athena, the goddess of wisdom and warcraft, he received a shield that was shiny as a mirror; and from Hermes, the messenger god, he was given a pair of winged sandals so he could fly through the air with ease.

Perseus heard about how Andromeda had been chained to a rock to be sacrificed to Cetus, the sea monster, and he vowed to save her. He first went to the cave of Medusa (the gorgon who had snakes for hair and turned you to stone with just one look). Medu-sa's reflection, although still ugly, wouldn't turn you to stone, so Perseus used Athena's mirror-like shield to warn him of Medusa's approach. When she got close enough, Per-seus closed his eyes and cut her head off with Hephaestus's sword. He put the severed head in a bag and flew on his winged sandals to Andromeda's rescue.

Perseus flew as fast as he could and made it just in time to see the fair Princess Andromeda about to be devoured by the sea monster. He yelled down to Andromeda to close her eyes, then he closed his eyes tight and pulled out Medusa's bloody, severed head. He showed the head to the sea monster who instantly turned to stone and fell to the bottom of the ocean, never to be seen again. Perseus unchained Andromeda, and they lived happily ever after.

The constellation Perseus is much more indistinct than his daring story might suggest. His outline is a squiggle of semi-bright stars that do not stand out on their own. With a little imagination you can picture a head, legs bent at the knees, an upraised sword in one hand, and an outstretched arm. At times, especially in early fall, the outline of Perseus looks a little like a stylized letter K.

## HOW TO FIND IT

**1** When in doubt, locate Perseus by first finding Andromeda. Locate the Big Dip-per, then connect the dots between the dipper's stars Merak and Dubhe.

**2** Continue on to Polaris, and then to Cassi-opeia's star, Caph. Keep going, make one more hop, and you will come to Androm-eda's head star, Alpheratz.

**3** From Alpheratz, the top of Andromeda's A-shaped outline, travel down to her feet (the stars at the bottom of the "A").

**4** Keep going in that direction and you will run into Perseus. Throughout nightly and seasonal journeys across the heavens, Andromeda's hero, the constellation Per-seus, remains at her feet and follows her footsteps across the sky.

## 77 | ALGOL

### The Ghoul

| CLASSIFICATION: **STAR** | VISIBILITY: **DIFFICULT** |

To save Andromeda, Perseus showed Medusa's bloody head to the sea monster and turned him to stone. The star in the Perseus constellation named Algol marks the pivotal point in this epic battle. It is the star that was believed to represent the head of Medusa. And it wasn't just the Greeks who thought that this was a monstrous star. Ancient stargazers from around the world noticed something strange about Algol. Sometimes Algol was bright, and at other times it was dim.

Whatever was happening to Algol, these changes were interpreted as bad omens. All the other stars seemed so steady and dependable that Algol's fluctuation of light scared ancient astronomers around the world.

The word *ghoul* comes from the Arabic name for this star. Other cultures referred to Algol by such colorful names as "Satan's Head," "Demon Star," and "Piled-Up Corpses."

In reality Algol is two stars that revolve around each other. Earth just happens to be in the prime position to see the stars eclipse each other at regular intervals. When one star eclipses the other, a lot of light pointed toward Earth is blocked out. You can see a noticeable dip in starlight about every 2.86 days.

### HOW TO FIND IT

**1** Algol (as Medusa's head) is the semi-bright star at the end of Perseus's arm. He is holding it out to his right and pointing it at the head of the sea monster constellation Cetus that, as you'll learn in the following entry, swims in the sky farther to the south.

**2** If you can picture Perseus facing you, Algol is in his left hand. It can be a challenge to find Algol when it dims, so if you detect a change in the outline of Perseus from night to night and one star seems to be missing, the culprit is the star Algol.

"We look pretty vulnerable
in the darkness of space."

Alan Shepard, astronaut

# CETUS

*The Sea Monster*

CLASSIFICATION: **CONSTELLATION** | VISIBILITY: **MODERATE**

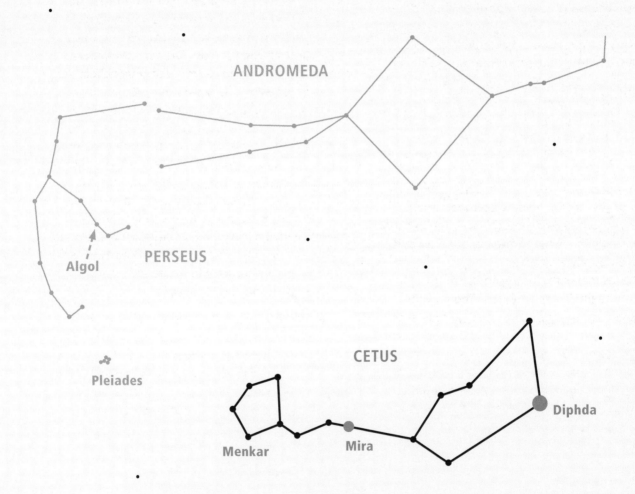

ANDROMEDA

PERSEUS

Algol

Pleiades

CETUS

Menkar

Mira

Diphda

In Greek mythology Cetus represents the sea monster that the god of the sea, Poseidon, could dispatch whenever and wherever he wanted individual humans to feel his wrath. In this case Poseidon sent Cetus to consume the fair maiden Andromeda in order to punish her family's vanity.

Andromeda was chained to the rock in the sea and soon Cetus emerged from the briny deep. Just as Cetus was about to chow down, who came flying along on his winged sandals? Perseus, to the rescue! When he arrived on the scene, he looked down to see the maiden in distress. Without a moment to lose, he yelled, "Andromeda! Close your eyes!" Perseus closed his eyes, reached in his bag, and pulled out Medusa's bloody, severed head and showed it to the sea monster. Cetus took one look at it and instantly turned into stone, cracked of his own monstrous weight, and sunk to the bottom of the ocean, never to be seen again. Perseus unchained Andromeda, they fell instantly and madly in love, and the couple flew away to live happily ever after.

Cetus marks the easternmost point of a watery realm in the heavens. The ancient Greeks designated several watery constellations in the fall sky, including Aquarius, the Water Bearer; Pisces, the Fish; Capricornus, the Sea Goat; and Piscis Australis, the Southern Fish. These join other water-loving summer constellations described earlier—Delphinus, the Dolphin, and Cygnus, the Swan.

## HOW TO FIND IT

**1** Set apart from the rest of the constellations in the fall sky, Cetus, the Sea Monster, can be a challenge to locate. It has several semi-bright stars, but only one, named Diphda, really stands out. If that wasn't bad enough, the outline of his stars resembles a recliner chair with a lumpy headrest more than they do a vicious sea monster. To locate him, first find Andromeda by connecting the dots between the Big Dipper's stars Merak and Dubhe. Continue on to Polaris, and then to Cassiopeia's star Caph. Keep going, make one more hop, and you will come to Andromeda's head star, Alpheratz. Go down her legs until you reach Perseus, and then imagine his outstretched arm with the star Algol pointing to the right. Travel 30 degrees to the right from Algol and you will meet a faint ring of stars marking Cetus's head. It's a large constellation, and Cetus's tail star, Diphda, lies another 40 degrees to the south of his head.

**2** You can start to see Cetus rising in the southeast after sunset in **OCTOBER**. In **DECEMBER**, Cetus sits about halfway up in the southern sky, and he will hang around until mid-**FEBRUARY** in the southwestern sky after dark.

# 79 | DIPHDA

## *The Frog*

| CLASSIFICATION: **STAR** | VISIBILITY: **MODERATE** |

Diphda, the brightest star in Cetus, the Sea Monster, is about 96 light-years away from Earth and is orange in color. Since the first millennium A.D. it has borne two different names, both from Arab astronomers: Deneb Kaitos and Diphda. Deneb Kaitos makes the most sense since it means "southern tail." However, in 2016 the International Astronomical Union (outer space's decision-making body) selected Diphda to be its official name, which enigmatically translates as "second frog." Beyond the name there is neither a surviving legend to describe the frog's adventures nor an explanation as to why a frog would be hanging out with a sea monster.

### HOW TO FIND IT

**1** One of the best months to look for Cetus is in **OCTOBER** when he is low in the east-southeastern sky. Diphda will rise in between two distinct objects in the fall sky: the Seven Sisters star cluster and the bright star Fomalhaut in the constellation Piscis Australis. The Seven Sisters will be about 60 degrees to the left of Diphda, and Fomalhaut will be about 25 degrees to the right. For a while in **OCTOBER** these three objects will be about the same altitude above the horizon and make a very long lineup.

**2** As **FALL** turns to **WINTER** you can watch Diphda arc from low in the southeastern sky to higher up in the south where it will reside 30–40 degrees above the horizon on **DECEMBER** evenings. By **JANUARY**, Cetus goes tail first toward the southwestern horizon, and Diphda will be the lowest star in the constellation— only 10–15 degrees above the ground.

# 80 | MIRA

## *The Wonderful*

CLASSIFICATION: **STAR** | VISIBILITY: **DIFFICULT**

Mira is a giant red star 300–400 light-years away in the constellation Cetus. Normally, Mira isn't visible to the naked eye. But for a few days every year Mira shines significantly brighter. It has acquired the nickname "the Wonderful" by those who first witnessed her flare-up. Mira's greatest outburst brought her to 1,500 times her normal brightness. On average, a brighter Mira occurs every 332 days.

What's happening with Mira? It is another binary star system—two stars, with a smaller one orbiting the larger one. As the two stars age, the smaller star is actually stealing mass from the larger star, and this in turn makes the amount of starlight that Mira emits somewhat unstable.

NASA's Galaxy Evolution Explorer (GALEX) satellite saw that Mira is also casting aside mass as it flies through space at over 80 miles per second. It's leaving something like a comet's tail as it goes, and this tail is now more than 13 light-years long.

### HOW TO FIND IT

**1** Mira marks the heart of the constellation Cetus, but you can only see it with the naked eye during a flare-up. Since that only happens every 332 days, you may have to wait a while to spot it.

**2** The best tip to locating Mira's place in the sea monster constellation is to first find Cetus's two brightest stars that mark the extremes of this constellation: Diphda and Menkar. Diphda is the tail of the sea monster and Menkar marks his head. Those two stars are separated by 40 degrees of sky, and Mira is on the line between them.

**3** Start at Diphda, the brightest star in Cetus, and start traveling toward Menkar. Mira will be about 27 degrees from Diphda or about three-quarters of the way to Menkar. You probably won't see Mira in this search, but if you're viewing on the right day in its 332-day cycle, maybe you'll catch this wonderful star.

# 81

# PEGASUS

*The Flying Horse*

| CLASSIFICATION: **CONSTELLATION** | VISIBILITY: **EASY** |

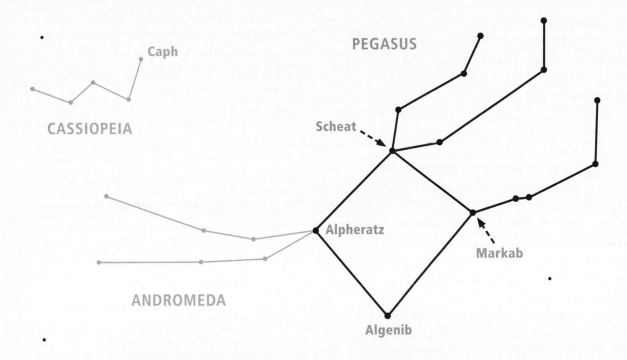

CASSIOPEIA

Caph

PEGASUS

Scheat

Markab

Alpheratz

ANDROMEDA

Algenib

In some versions of the Pegasus myth, Perseus rode this flying horse to save Andromeda from Cetus, the sea monster. However, most versions portray Pegasus arising after the dramatic climax in one of the strangest origin stories.

During Perseus's battle with Cetus, Medusa's head (which Perseus had cut off to turn the sea monster into stone) was still bleeding. Perseus had just killed her that morning, after all. When Perseus held out Medusa's head to the sea monster, some blood dripped out of her neck. When the blood hit the sea water a magical thing happened: Medusa's blood mixed with the water and turned into Pegasus, the Flying Horse. This was actually a common theme in Greek mythology—when a monster or god bled, something always sprang from it.

When you picture Pegasus in the sky, just imagine him flying upside down. Look for a big square or diamond shape of semi-bright stars. This is Pegasus's body. His head, mane, and front legs jut out from the westernmost parts of the square. You may imagine that you can see Pegasus's back legs toward the east, but remember those stars are actually part of Andromeda.

The four stars making up the Great Square of Pegasus are interesting sights in their own right. Starting from Alpheratz (also Andromeda's head) and heading clockwise around the square, you will come to Scheat, then Markab, and then Algenib. The names of these stars are Arabic in origin and are meant to help illustrate different body parts of the horse. *Scheat* means "shoulder" or "upper arm," *Markab* means "saddle," and *Algenib* means "the side" or "the wing."

The four stars in this square are all different colors and distances away. Alpheratz is blue-white in color and lies about 97 light-years from Earth, Scheat is a red star 196 light-years away, Markab is blueish-white and 133 light-years out, and Algenib is a deeper blue and a whopping 390 light-years away.

## HOW TO FIND IT

**1** The best way to find Pegasus is to do some star hopping. Start at the Big Dipper, and then connect the dots between the stars in the bowl of the spoon, Merak and Dubhe. Continue that line for another 30 degrees to get to Polaris, and then go another 30 degrees to reach Cassiopeia's star Caph. Keep going and make one more 30-degree hop. This brings you to Alpheratz, the head of Andromeda that also doubles as Pegasus's rear end. It's a long way to go (90 degrees, or one-quarter of the way around the sky), but that is the best way to find one corner of the Great Square of Pegasus.

**2** You may also recognize the Great Square of Pegasus by noting a lack of stars within the square. The Great Square contains surprisingly few naked-eye stars for such a large area of the sky.

**3** You can start to see Pegasus rise in the eastern sky just after sunset in **SEPTEMBER**. At that angle it will look like a baseball diamond and Andromeda's head star, Alpheratz, marks third base. In **NOVEMBER** and **DECEMBER**, Pegasus flies high in the southern sky (but still upside down). By **JANUARY** look for Pegasus in the western sky every evening. By mid-**FEBRUARY** you will catch a last glimpse of it as it sets until next year.

# 82

# PISCIS AUSTRALIS

*The Southern Fish*

CLASSIFICATION: **CONSTELLATION** | VISIBILITY: **MODERATE**

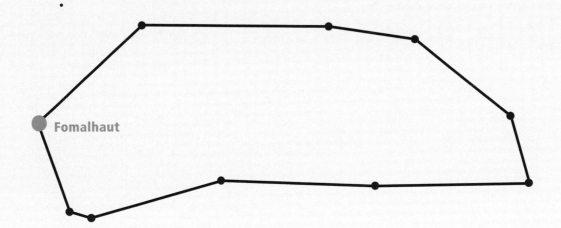

Fomalhaut

Piscis Australis (or Piscis Austrinus) is a little constellation that swims above the southern horizon during the fall. Piscis Australis is the Southern Fish, and the big daddy of the two fish that form the zodiac constellation Pisces.

Piscis Australis has one really bright star called Fomalhaut (pronounced *foam-a-lot*), which means "fish's mouth." The other stars in the constellation are so faint that none of them have common names. The outline of these fainter stars is often drawn in an oval shape. However, if you have a lot of imagination and you are viewing under a really dark sky, maybe you can reconnect the stars of Piscis Australis differently. See if you can form them into a line drawing of a goldfish made by a kindergartner.

## HOW TO FIND IT

**1** The best way to locate Piscis Australis is to find Fomalhaut, its brightest star. If you look in the southeastern sky at the beginning of **FALL**, it will be the only bright star in that part of the sky.

**2** By mid-**FALL** it shines low in the southern sky, and in **DECEMBER** it is low in the southwest. By **JANUARY**, Fomalhaut and Piscis Australis are lost in the sunset as the **WINTER** constellations arise. But each and every **FALL** evening, you can look for the Southern Fish with the lone bright star in its mouth.

# 83

# FOMALHAUT

## *The Fish's Mouth*

| CLASSIFICATION: **STAR** | VISIBILITY: **EASY** |

During the fall evenings one star stands alone in the southern sky: Fomalhaut. One translation of this star's name is "fish's mouth." But this star was also known to some ancient astronomers as the "first frog" that hopped across the sky ahead of the "second frog," the star Diphda, in the constellation Cetus.

There are no other first magnitude stars equal or brighter than Fomalhaut within 50 degrees of it. Its solitary location made a perfect landmark for measuring the fall sky. In addition, Fomalhaut is one of the closer bright stars to us at only 25 light-years away. Its proximity has made it a great target for astronomers. This star is so bright that astronomers decided to mask out its glare in order to see what's around it. When they did this, they found rings and rings of dust circling around Fomalhaut. Embedded in that dust was a massive planet, looking like a slow-moving lump around the fish's mouth. This was the first planet to be seen visually orbiting another star.

Some astronomers say that there aren't any green-colored stars, but this is open to individual interpretation. Fomalhaut twinkles blue, white, and sometimes green. Find Fomalhaut for yourself and see what you think.

### HOW TO FIND IT

**1** Fomalhaut, the lone star of the south, is the brightest star in the constellation Piscis Australis.

**2** Look for it rising in the **OCTOBER** evening skies above the southeastern horizon. In **NOVEMBER** it will be the only bright star low in the southern sky, and by **DECEMBER** and **JANUARY** it will set in the southwest not too long after the Sun goes down.

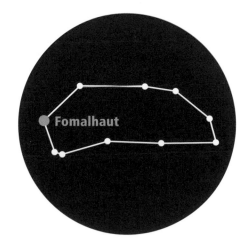

"All that we have accomplished in space—all that we may accomplish in days and years to come—we stand ready to share for the benefit of all mankind."

Lyndon B. Johnson, American president

# 84

# ARIES

## *The Ram*

| CLASSIFICATION: **CONSTELLATION** | VISIBILITY: **DIFFICULT** |
| --- | --- |

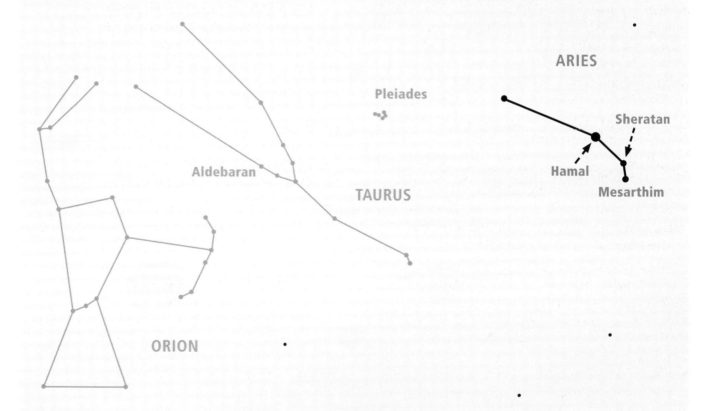

**ARIES**

Pleiades

**TAURUS**

Aldebaran

Sheratan

Hamal

Mesarthim

**ORION**

Aries is another zodiac constellation that helped many ancient civilizations recognize the changing seasons and alert them to the best time to plant and harvest crops. During the spring the Sun moved through Aries, signaling it was time to begin planting. And when Aries became visible at night early in the fall it was a signal that the harvest season had begun. As the constellation of agriculture it earned the designation Prince of the Zodiac.

For such a tiny constellation Aries features mightily in several Greek myths. One begins when a king, with a son named Phrixus and a daughter named Helle, leaves his wife and remarries a mean and nasty woman. The new wife is jealous of the children and plans to sacrifice them to the gods. At the last moment their biological mother sends a magical, winged, golden ram to rescue Phrixus and Helle to fly them away to the east. Unfortunately, the daughter falls off of the flying ram to her death in an area still called the Hellespont (named in her honor). Phrixus lands safely, sacrifices the ram to Zeus, and gives the Golden Fleece to another king named Aeetes (whose daughter Phrixus eventually marries).

Years later a Greek hero named Jason gathers a group of explorers. Calling themselves the Argonauts, they set out to reclaim the Golden Fleece. The Argonauts, which included Hercules, Orpheus, Castor, and Pollux, among others, have a series of epic adventures and ultimately succeed in their quest for the Golden Fleece.

The stars of Aries don't look anything like a ram, however. The outline of the three major stars in Aries resemble the slight curve of a boomerang. Only one star in the constellation is significantly bright: Hamal. The others are second, third, and fourth magnitude stars that are difficult or impossible to see from skies with light pollution.

Hamal lies 66 light-years away and is orange in color. Its name is an Arabic word meaning "leader of the sheep." The star has also had many other cultural names, such as "the Ram's Eye" and "the Horn Star." The Akkadians believed Hamal to be the king named Aloros, who was one of ten bright stars in the path of the zodiac. These kings reigned supreme over the world just after creation. And finally, eight temples of ancient Greece were dedicated to this star—the Greeks do value their sheep.

## HOW TO FIND IT

1. Aries is extremely difficult to find because it has only three stars of note: Hamal, Sheratan, and Mesarthim. But other stars and constellations can help you locate the ram's place in the sky.

2. First find Orion's Belt. Then connect the dots of the belt and continue that line to the right or east for 23 degrees. This view will take you under the ruddy star Aldebaran, which is in Taurus, the Bull. Keep going another 12 degrees until you reach a tight, bright group of stars called the Pleiades. Continue that line, arcing it a little to the right, and after another 23 degrees you will run into Aries's brighter star, Hamal.

# Part III.

# Beyond Stargazing

So far we've discussed how to view the Sun safely, what to look for on the Moon, the motions of the five naked-eye planets, and how to identify dozens of stars, constellations, and several deep space objects. What else can you see with the naked eye on a clear night? Lots.

If you spend any time stargazing, every so often you'll definitely notice a slow moving, nonblinking light crossing the heavens. What you're seeing is one of the 4,000 man-made satellites that circle the Earth every night. Launched into space by rockets, these satellites aid global communications, conduct scientific research, and sometimes undertake secret missions. Satellites shine only from reflected sunlight. While you are standing in the dark on the ground, the satellites are high enough to be lit by the Sun. The sunlight reflects off their surfaces, and you can see them against the darker background, making them look like little stars sailing through the blackness of space.

Or, on any given night you might also see a meteor streaking across the sky. Most people call them shooting stars, but their origins and fiery deaths are a lot closer to home than the distant stars. Meteors are best seen during annual meteor showers when you can see several dozen crossing the sky per hour. The greatest showers, which occur once in a generation, are called meteor storms. You can see so many shooting stars pass overhead that you'll think the sky is actually falling.

Finally, you can observe some of the greatest astronomical events with the naked eye. Some of these sights are quiet and subtle, like observing the Milky Way or zodiacal light. Some require timing and patience, like observing planetary conjunctions and occultations. And others, like auroras and eclipses, will blow your mind. You have to be at the right place at the right time (and perhaps have a little luck), but these special heavenly events will definitely wow you.

# Catching Satellites

It's not a bird, not a plane, and not a meteor. That steadily moving light in the sky is probably a satellite. How do you know for sure? There are some surefire ways to tell. If you see a blinking light in the sky, it is probably a plane. Satellites almost always shine with a steady light. They may gradually brighten or dim, but they do not blink.

Satellites also never suddenly change direction. They always travel in a steady arc across the sky. They are fast but not flashy, and the speed of a satellite depends on its altitude above Earth. Most satellites take 4–6 minutes to cross the sky from horizon to horizon. If you see something moving much more quickly (if it crosses the entire sky in only a few seconds), know that you probably saw a meteor/shooting star.

The brightest satellite and the easiest to spot is the International Space Station (ISS). With a little practice you might be able to find the Hubble Space Telescope and the X-37B (a super-secret unmanned space plane also known as an Orbital Test Vehicle, or OTV).

To know when to look for certain satellites, websites like www.heavens-above.com and apps like Satellite Tracker can help. Once these tools know your location, they can tell you the exact time when, for example, the ISS will pass overhead, in which direction you should look, how high in the sky it will fly, and how bright it will appear. With this prior knowledge you can amaze your friends by taking them outside and showing them a satellite cutting across the sky in the exact position that you predicted. They'll think you're a rocket scientist!

What is 356 feet long, 239 feet wide, weighs nearly 1 million pounds, and circles Earth every 92 minutes? It is the International Space Station (ISS), the largest satellite humans have ever constructed in space.

The ISS is a joint effort between the United States, Russia, and several other countries. It went up into space, piece by piece, starting in 1998. It is so large that it can accommodate a crew of six full-time residents as well as entertain visiting guests (the replacement crew). The purpose of the ISS is to provide a laboratory in space to test out new technologies that can help us better understand Earth and prepare us for a longer space mission to Mars and beyond.

Like most satellites the ISS is surprisingly close to Earth. It orbits between 205 and 260 miles up, which is roughly the distance from Cleveland to Cincinnati. To stay in orbit, the ISS must travel at more than 17,000 miles per hour. Every once in a while rocket boosters bump it into a higher orbit so that this object the size of a football field will not fall to Earth.

Astronauts aboard the ISS get an amazing view of Earth and the sky. Since they circle Earth so quickly, they experience one sunrise and one sunset about every 90 minutes.

## HOW TO FIND IT

**1** From Earth the ISS looks like an exceptionally bright, slowly moving star. Like other satellites (and unlike stars), it should not blink. If you were to observe its entire path across the sky, it would take about 5–6 minutes for it to travel from horizon to horizon.

**2** Because of its massive size the ISS can shine like a dazzlingly bright beacon. Even at its minimum brightness, it shines with a glow that is similar to a first magnitude star. When it flies directly overhead or reflects the maximum amount of sunshine toward you, the ISS can be almost as bright as the planet Venus. The ISS can fly over your town from nearly any direction and at nearly any time. Satellite tracking apps and websites can alert you when and where to look. When you see it, be sure to wave at the astronauts inside that are about 250 miles above you.

| CLASSIFICATION: **SPACE STATION (MANNED SATELLITE)** | VISIBILITY: **EASY** |

You may be surprised to hear that a secret, unmanned, reusable US Air Force spacecraft, an Orbital Test Vehicle (OTV) nicknamed the X-37B, circles the globe from time to time. It was first observed by amateur astronomers in 2010, when no one other than the US military knew about it. What is the OTV doing? The Air Force has never said, so no one is really sure.

The OTV is launched into space by a rocket, discards the rocket, and then circles Earth. After months or even years, it then reenters Earth's atmosphere and lands like a plane at Vandenberg Air Force Base in California or at NASA's Kennedy Space Center in Florida. A version of this spacecraft/satellite has completed five missions of various lengths. The last mission, OTV-5, was in orbit between September 7, 2017, and October 27, 2019.

## HOW TO FIND IT

**1** Will the OTV fly again? The sixth mission, OTV-6, may launch sometime in 2020. If it does, you can be sure that amateur astronomers will continue to keep tabs on it and publish its path online. So even if you don't know what this satellite is doing, you can know where and when it flies.

**2** The OTV can shine with the equivalent brightness of a first, second, or third magnitude star. At its brightest, it is as bright as the stars Vega and Arcturus, but when the Sun's angle is not optimal, it can appear fainter than the North Star.

# 86

# OTV

## X-37B or the Secret Spaceship

CLASSIFICATION: **SPACECRAFT**   |   VISIBILITY: **DIFFICULT**

On some evenings or early mornings you can spot the Hubble Space Telescope (HST), one of the greatest scientific instruments humans have ever created. This satellite revolves around Earth at an altitude of about 336 miles. From that lofty perch it takes about 95 minutes to circle the globe. Its brightness usually hovers between second and third magnitude—about the brightness of the fainter stars in the Big Dipper—but it is definitely visible to the naked eye.

The space shuttle *Discovery* ferried the Hubble Telescope into space in 1990. Astronauts took it out of the cargo bay, released it into space, and sent it forth to circle the globe ever since. From its vantage point in space HST has made countless astronomical discoveries and has taken the most iconic space pictures of our generation. The telescope sports an 8-foot-diameter mirror that can detect distant galaxies and stars with a thirtieth magnitude brightness, which is almost 4 billion times fainter than what the naked eye can see.

Time is winding down for Hubble. After 30 years in space it is still working well. However, it is only a matter of time before something on it breaks. After that happens it will make a slow descent into the atmosphere and burn up. Meanwhile, Hubble's replacement, the James Webb Space Telescope (JWST), after countless delays, should launch in the 2020s.

## HOW TO FIND IT

**1** Spotting the Hubble Space Telescope can be a real challenge. Most of the time you would not notice it when it passes overhead. It moves like an ordinary satellite but only shines with the equivalent brightness of a third magnitude star. That means it is only about half as bright as the North Star.

**2** Your best bet is to consult satellite tracking apps or websites that can tell you exactly when and where to look for it.

# 87

# HUBBLE SPACE TELESCOPE

CLASSIFICATION: **SATELLITE** | VISIBILITY: **DIFFICULT**

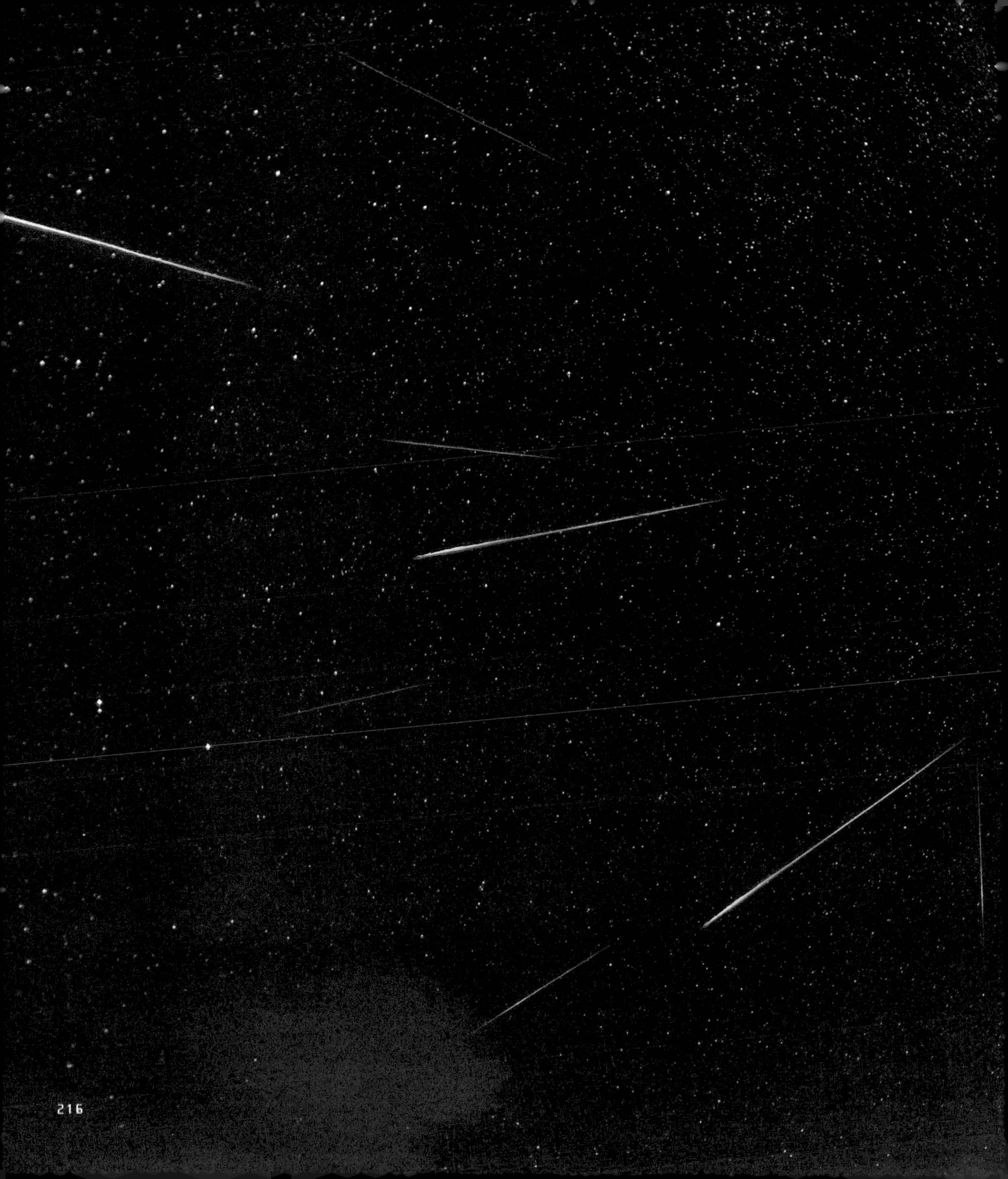
216

Head outside on any clear night, sit back in a comfortable chair with a warm beverage, and watch the light show up above. Chances are, if you stay out long enough, you will see a meteor streak across the sky. Also called shooting stars, meteors are objects falling through Earth's atmosphere.

A meteor exists for only a brief moment; most are only visible for a few seconds. And, despite their occasional brilliance, most meteors are incredibly small. They are usually about the size of a grain of sand and rapidly burn up before hitting the ground.

When you see meteors, they may look close to you but they are not. They light up when they are about 40–50 miles above Earth in the upper atmosphere. Meteors blaze because they are decelerating from tens of thousands of miles per hour to hundreds of miles per hour. That rapid deceleration transmits into heat and causes the air around the meteor to glow. Once the meteor is lower in the sky and has stopped this rapid deceleration, it stops shining.

Meteor showers occur at predictable dates each year when Earth slams into a swarm of space debris left behind by passing comets or asteroids. To maximize your meteor shower viewing experience, get out of the cities and away from lights. The darker the sky, the more meteors you will see. The best times to view meteor showers are generally between 2 a.m. and 5 a.m. You can see some early shooting stars around midnight, but the later you stay up the better. Once you start to detect hints of sunrise coming in the east, the number of meteors you will see will greatly diminish. So let's take a look at some shooting stars!

Really bright meteors are called fireballs. They can glow all sorts of colors including white, blue, and green. Really dramatic fireballs can break up into multiple pieces and become multiple meteors streaking across the sky.

Most fireballs were originally pieces of asteroid fragments. Astronomers have charted more than 700,000 asteroids in our solar system. These irregularly shaped chunks of metal, rock, and dust circle our Sun mainly between the orbits of Mars and Jupiter. But some very small ones can come close to Earth and even run into us.

And that is exactly what happened on February 15, 2013, when a huge meteor streaked across the sky over Chelyabinsk, Russia. For a brief moment it shone brighter than the Sun and cast stark shadows. At first it did not make a sound, but about 2 minutes later a sonic boom shattered windows and even knocked people over. The speed of light is 186,000 miles per second while the speed of sound is only 767 miles per hour. That difference in speed accounted for the lag time between seeing it and hearing it. If you see an incredibly bright fireball explode above the ground, stay away from windows. A sonic boom may strike a few minutes later and blow out the window in front of you. However, not a single person in the past 100 years has been struck and killed by a meteorite or falling space debris, so there is nothing to fear when you are conducting your meteor watch.

## HOW TO FIND IT

1. Every day between 10 and 100 tons of material falls into the atmosphere from outer space. While you can sometimes see a chunk of man-made debris entering the atmosphere and turning into a fireball, the vast majority of it burns up and/or falls into the ocean.

2. Observing a fireball is a rare and random event that astronomers cannot yet predict.

# 88

# FIREBALLS

*Great Balls of Fire*

CLASSIFICATION: **METEOR** | VISIBILITY: **MODERATE**

After centuries of watching the skies, astronomers figured out that more meteors fall on certain days of the year than on others. These events are called annual meteor showers and several of them provide consistently impressive celestial performances.

When a comet passes near the Sun, it sheds icy material to form a long tail. Tail particles are mainly tiny pieces of ice and dust that can accumulate in great clouds of debris along the comet's orbit. Earth slams into this cometary debris on a yearly basis, and the result is a predictable schedule of meteor showers. And when these clouds intersect with Earth, get ready.

The most impressive annual meteor showers are the Perseids, Orionids, Leonids, and Geminids. Meteor showers get their names from the constellation from which the shooting stars seem to radiate. The meteors don't actually fall from the stars in Perseus, Orion, Leo, and Gemini: They just appear to do that. They actually ignite in Earth's atmosphere, about 40 miles up.

Astronomers know exactly where these meteors come from. The Perseids are remnants of the comet Swift-Tuttle, the Orionids come from Halley's Comet, the Leonids stream from the tail of comet Tempel-Tuttle, and the Geminids originate from pieces broken off an asteroid named 3200 Phaethon.

The best times to view meteor showers are generally between 2 a.m. and 5 a.m. You can see some early shooting stars around midnight, but the later you stay up the better. Don't use telescopes or binoculars: Your naked eye is all you need. Take in as much of the sky as possible to catch as many stray streaks as you can.

Get away from the lights of the city and find a dark sky location where you can detect even the faintest meteors. Avoid meteor showers that peak during a Full Moon, because the moonlight will wash out all but the brightest meteors. And keep your expectations low. While some astronomers predict that you'll see dozens to hundreds of meteors per hour, it is more realistic to expect 10–20 per hour.

## HOW TO FIND IT

1. The Perseids are the meteor shower of summer, peaking each year on **AUGUST 12 OR 13**. You want to generally face in the direction of the constellation Perseus, which rises in the northeastern sky at midnight and shifts high in the south later at night, but meteors can streak from any direction at any time between midnight and dawn. Be patient and vigilant.

2. The most famous comet in the sky, Halley's Comet, is responsible for the Orionid meteor shower. The Orionids peak every year around **OCTOBER 21**. Although you can't see Halley's Comet in all its glory until it passes near Earth again in 2061, you can still see its remnants from previous flybys in the sky every October. Face the constellation Orion on October 21 between 1 a.m. and 5 a.m. to see the maximum number of Orionids.

# 89

# THE PERSEIDS

## Orionids, Leonids, and Geminids

| CLASSIFICATION: **METEOR SHOWERS** | VISIBILITY: **MODERATE** |

**3** The Leonids produce reliably good meteor showers every **NOVEMBER 17 OR 18**. To find them, lie back and search the skies around the constellation Leo as it comes up in the eastern sky after midnight and slowly arcs to the south later at night. If you watch all night until the break of dawn, you may see dozens of meteors per hour.

**4** One annual meteor shower that has become increasingly visible in the twenty-first century is the Geminids. Every **DECEMBER 13 OR 14**, scan the sky around the constellation Gemini between 2 a.m. and 5 a.m. to see the peak of this unusual meteor shower. The Geminids are not cometary in origin—they actually come from an asteroid called 3200 Phaethon, a rocky and icy body circling the Sun. As it regularly flies close to Earth, 3200 Phaethon leaves bits of rock and ice behind. Every December, Earth flies through these asteroid parts and stargazers are treated to a solid meteor shower.

"When we contemplate the whole globe as one great dewdrop, striped and dotted with continents and islands, flying through space with other stars all singing and shining together as one, the whole universe appears as an infinite storm of beauty."

John Muir, naturalist and author

METEOR STORMS

CLASSIFICATION: **MOTHER OF ALL METEOR SHOWERS** | VISIBILITY: **DIFFICULT**

Better than a shooting star, better than a meteor shower, a meteor storm is a once-in-a-lifetime experience. Like meteor showers, meteor storms originate from the leftover debris of comets. While during normal meteor showers Earth plows through some of this debris, during a meteor storm our planet crosses paths with swarms of denser cosmic material. During a meteor storm you can observe about one hundred meteors per minute raining down from all directions. This is the heavenly fireworks show that made witnesses, both ancient and modern, truly think the sky was falling.

The annual Leonid meteor showers have produced the most memorable storms in recent history. In a fiery display in November 2001, observers counted more than 2,000 Leonid meteors per hour. But that was nothing compared to the Leonid meteor storm of 1833 where some astronomers estimated between 100,000 and 240,000 meteors streaked across the sky in one night. Astronomy writer Agnes Clerke described the scene: "On the night of November 12–13, 1833, a tempest of falling stars broke over the Earth....The sky was scored in every direction with shipping tracks and illuminated with majestic fireballs."

## HOW TO FIND IT

**1** Unfortunately, meteor showers rarely turn into meteor storms. Astronomers have linked the Leonids to the debris from comet Tempel-Tuttle, which circles the Sun every 33 years. And almost every 33 years Earth slams into thicker remnants from this comet's tail. Since we know Earth's orbit and Tempel-Tuttle's orbit in great detail, astronomers try to predict which years could put on a better meteor show than others. After a superb display in 1966, astronomers figured that the next possible meteor storm would occur in 1999. It didn't happen that year. Instead, the big storm came 2 years later, in 2001.

**2** That's the thing with meteors: They are predictably unpredictable. Assume nothing when you try to look for shooting stars. The key is to observe the sky between 2 a.m. and 5 a.m. during the peak of each meteor shower. You will most likely see just a few, but if you're lucky and a meteor storm breaks out, don't panic. Just sit back, "Ooh" and "Aah," and enjoy the grand show.

# More Astronomical Events to See with the Naked Eye

There is a loose organization of people from all walks of life who make it their purpose in life to see eclipses. They save money and plan vacations around astronomical events. They travel to the far corners of the globe for a glimpse of the greatest shows from Earth. They are called eclipse chasers.

Witnessing a total solar eclipse is a powerful experience. When the Moon covers the entire Sun and turns the sky dark in an instant, it does something to people. They yell, they cry, they hoot and holler. Even rational individuals are temporarily transformed into giddy little children under the influence of the total solar eclipse. Eclipse chasers relish this rush, and afterward they have a hard time thinking of anything else except seeing another.

This is the power of the universe. Some astronomical events like total solar eclipses are just downright awesome. And when the northern lights ignite above you in the upper atmosphere or a dynamic comet haunts the night sky, you will remember it for the rest of your life. When the weather is perfect, light pollution is at a minimum, and the heavens align, the universe can put on the most incredible light show. And with a lot of planning and a little luck you can share in the most spectacular, can't-miss, you-gotta-see-to-believe astronomical events in the daytime and nighttime sky. Put these on your bucket list to behold sometime in your life.

In this final section I challenge you to take a trip to the country and get away from the city lights to see the Milky Way and zodiacal light on a moonless night. Prepare for perfect celestial lineups like lunar eclipses, planetary conjunctions, and lunar occultations. Get surprised by the sudden beauty of a great comet or mesmerized by the appearance of the aurora borealis (or northern lights). And chase a total solar eclipse around the world. These final things to see in the sky range from super cool to life changing, and like everything described in this book, you can experience them all with just your naked eye.

Have you ever seen a bright "star" near the Moon on a dark evening? If you took special note of the scene, you'd see that "star" was probably a planet. Every month the Moon slides past all five naked-eye planets (Mercury, Venus, Mars, Jupiter, and Saturn). Our solar system is like a flat disc and the planets circle the Sun on nearly the same level plane. From Earth this disc looks like a curved line across the sky, called the ecliptic. Since the Sun and Moon inhabit this line, lunar and solar eclipses can only occur on the ecliptic. The ecliptic cuts through some very famous constellations such as Sagittarius, Leo, Taurus, and Virgo. These constellations, along with eight others, form the twelve zodiac constellations and help astronomers map out the movement of the naked-eye planets. When someone says, "Jupiter is in Leo," they mean that the planet Jupiter can be found in front of Leo's star pattern.

Since you can find all of the planets on or near the ecliptic, they often line up. You may have heard doomsday prophecies like, "Watch out when the planets align!" The truth is, the planets align all the time! That's just what they do.

When a planet and the Moon (or two or more planets) appear exceptionally close to each other, astronomers call that a conjunction. Any combination of the planets and Moon make an impressive sight in the sky. Although two bright lights standing side by side is an inspiring sight to behold, some conjunctions really make people stand up and take notice.

## HOW TO FIND IT

**1** It is an amazing sight when the brightest planets, Venus and Jupiter, have a conjunction with the Moon or each other. A Venus-Jupiter conjunction happens about every year but some are more intimate than others. Good conjunctions bring the two planets within $\frac{1}{6}$ of a degree. Sometimes they are so close you can barely tell them apart.

**2** The most astronomical planetary conjunction happens when you can see all five naked-eye planets at one time. In May of 2002, Mercury, Venus, Mars, Jupiter, and Saturn were all within 33 degrees of each other. They didn't perfectly line up but instead looked like a planet clump. The next time these five planets will gather so close together will be in 2040, but occasionally you can see them stretched out from horizon to horizon and aligning in the night. To find when the best conjunctions will occur, browse websites that offer "astronomical highlights" for the year ahead. Each month should offer the dates when the Moon cozies up to a planet, and be on the lookout for the rarer occurrence when two to five planets come together in cosmological conjunctions.

# 91

# PLANETARY CONJUNCTION

A lunar occultation occurs when the much closer Moon seems to block out the light of a much farther planet or star. It's like a super eclipse. As the Moon orbits Earth it seems to wander across the background stars very quickly (moving about 13 degrees per night). The path the Moon takes is very similar to the ecliptic (the imaginary line where the planets hang out). So the Moon regularly passes through the stars in the zodiac constellations in addition to coming close to the five naked-eye planets.

The following are four really bright stars that are close enough to the ecliptic to be occasionally occulted:

- Aldebaran in Taurus
- Regulus in Leo
- Spica in Virgo
- Antares in Scorpius

Even though the Moon passes through the zodiac constellations quickly and repeatedly, it is still very rare for the Moon to occult a bright star or planet. Additionally, these alignments are so precise that they are not visible from everywhere on Earth. You have to be at the right place, at the right time, to see a lunar occultation.

## HOW TO FIND IT

**1** Every year some astronomy websites list all of the lunar occultations coming up and where you have to be to experience them. Simply search on the Internet for "lunar occultations for" and then type in the current year and you'll find a host of solid links.

**2** On the day of a lunar occultation, this is what will happen: When you look at the Moon, there will be a bright star or planet just to the Moon's left. As the minutes pass, you can't see the Moon moving, but the distance between the two objects will noticeably shrink. When it looks like the star is touching the Moon, get ready because in one moment the star will be there—and the next moment it will be gone, covered by the Moon. Now you see it, now you don't. The Moon has no atmosphere, so when the rocky surface of the Moon covers it, the occultation happens almost instantaneously.

**3** When the Moon occults a star or planet, it does so for no more than about an hour. Not only is it amazing to witness the light of a star or planet suddenly disappear, but you can also see it blink back on again as the Moon uncovers it and continues on its journey around Earth.

# 92

# LUNAR OCCULTATION

**CLASSIFICATION: RARE LUNAR EVENT** | **VISIBILITY: DIFFICULT**

For city dwellers the Milky Way is completely invisible. City lights are so bright that light pollution robs many people in the United States of the chance to see a dark starry sky. Because of urban light pollution many people have never seen the Milky Way. But the Milky Way is still there, waiting for you to observe it. And when you see it, it is spectacular!

At first glance the Milky Way looks like a high, thin cloud stretched out in a long line across the sky. It may even seem that your eyes are playing tricks with you. As your eyes adjust to the darkness it looks like someone spilled milk in the heavens. But what you are really seeing is the flat disc of our galaxy. The "milk" is actually billions and billions of stars shining at you from tremendous distances. The combined light reaches your eyes as a diffuse cloudy structure—brighter patches where you can discern individual stars interspersed with dark dust lanes that seem to split the Milky Way into channels. Best of all, no telescope is needed to experience it.

## HOW TO FIND IT

**1** It may take a little extra effort (and a small road trip) to see the Milky Way stretch magnificently under a truly dark sky. Drive out of the city on a clear night and you will discover a wealth of stars beyond your imagination. On your next **SUMMER** vacation make an effort to view the stars from a national park or a really isolated area. Investigate the least light-polluted places in America and plan a visit. Go there and you'll see how the night sky is supposed to be.

**2** You can see the Milky Way all year round, but some seasons provide better viewing than others. Whenever you see the constellation Cassiopeia, the Milky Way is there too. The queen seems embedded in a multitude of stars. During the **WINTER** months the Milky Way runs near the constellation Orion and between his dogs, Canis Major and Canis Minor. Later in the **SUMMER** it arches high in the sky near Cygnus, the Swan, and Aquila, the Eagle, and stretches down to Sagittarius and Scorpius.

# 93

# MILKY WAY

**CLASSIFICATION:** OUR GALAXY | **VISIBILITY:** MODERATE

Almost 1,000 years ago the Persian poet Omar Khayyam in his book of poetry, *The Rubaiyat*, wrote his most famous line: "A jug of wine, a loaf of bread and thou beside me singing in the wilderness." But elsewhere Khayyam made a poetic allusion to a mysterious false dawn. It took astronomers centuries to figure out what he was writing about, and it makes a rare treat to witness. The event is called the zodiacal light.

There are trillions of dust particles that populate our inner solar system. These particles are incredibly small and are spread out over millions of square miles. But when the Sun strikes them, their combined mass can reflect a diffuse light toward Earth. The reflected sunlight bouncing off the cosmic cloud of comet and asteroid debris creates the zodiacal light that is barely perceptible to the naked eye. This warm, subtle zodiacal light fooled late night stargazers into thinking the dawn was about to break even though daybreak was still hours away.

## HOW TO FIND IT

**1** To see this zodiacal light, you will need to view the sky during or near a New Moon and be far from city lights. If you can see the Milky Way clearly, then you may have a chance to see it. Face east a few hours before sunrise. The zodiacal light will look like a dim, cone-shaped patch of powdered sugar that extends from the horizon to about one-third of the way up in the sky. It's also called the false dawn because it will give you the illusion that dawn is about to break, even though sunrise is still hours away.

**2** The zodiacal light can also be seen in the evening, a "false dusk," which you can see toward the western horizon a few hours after sunset. Both phenomena are caused by the same thing: trillions of dust particles circling around the Sun and reflecting its light. You can observe both events during any night around the New Moon. However, it is easier to see the morning zodiacal light in the late **FALL** when the bulk of the particles appear higher in the sky, while **SPRING** is the optimal viewing season for the evening zodiacal light when the light reflecting off this space dust is higher above the western horizon.

**3** This extremely subtle light show may not bombard your senses, but once you see it, and realize what you are seeing across millions of miles of space, it should evoke a quiet wow moment.

CLASSIFICATION: **ILLUMINATED SPACE DUST** | VISIBILITY: **DIFFICULT**

Comets are small chunks of ice only a few miles wide that circle the Sun in long, looping orbits. For most of their lives, they dwell at incredible distances from the Sun and are locked and frozen. But for a few months during every orbit, they may swing so close to the Sun that they heat up. Ices turn to gases and geysers of material erupt from the nucleus of the comet to form a bright head and a long tail behind it.

About once every decade stargazers are treated to a bright comet that lights up the night sky. The sight of a comet truly inspires us with wonder and awe. There is something about that fuzzy little visitor with the long tail that excites the imagination. Where did the comet come from? Where will it travel to? How long has it been orbiting the Sun? And how many times has it swung by Earth?

There is something unearthly and maybe a little unsettling about a bright comet. It often takes weeks or months to brighten up significantly, so each night you see it, it gets a little bigger. No wonder the ancients feared comets. After all, they looked like they were coming right at them!

A comet seems to just hang in the sky with a fuzzy bright head called the coma and long wispy strands of gas jetting off to form a tail. They can haunt the night sky for weeks at a time, brighten unexpectedly, or quickly fade from sight. Although comets travel around the Sun in a predictable path, even astronomers don't know what they will look like or how dazzlingly bright, or boringly mediocre, they will become.

Although we have had dozens of comets swing by Earth's neighborhood in the twenty-first century, none of them have been truly spectacular. In fact, the last great comet visible to stargazers in the Northern Hemisphere was comet Hale-Bopp in 1997. This is an extraordinarily long drought. We're overdue for a comet of the century that will wow anyone and everyone who looks up.

## HOW TO FIND IT

1. A truly great comet is not only visible to the naked eye, but its tail covers 10, 20, even 30 degrees of the sky. When you walk outside, it immediately grabs your attention. It's difficult to look away. Unfortunately, astronomers are not expecting any known comets to brightly light up the night in the next few decades.

2. However, most great comets seem to pop up out of nowhere and catch us off guard. They are small, stray ice balls from the outer reaches of the solar system. And we often can't see them until they come closer to the Sun and cast their ethereal tails across the night sky. An optimistic astronomer might say, "Hopefully we will see one next year."

A GREAT COMET

CLASSIFICATION: **A COMET OF THE CENTURY** | VISIBILITY: **DIFFICULT**

When you step outside at night and look at the stars, what does it mean if the sky is green? What about if red and white streaks of light start swirling and dancing in waves and curtains? It's not the end of the world. It's the aurora borealis, the northern lights!

Pictures do not do this phenomenon justice. Whole swaths of the sky are enveloped in color and light. The weaker shows involve a green-colored fog rolling in from the north while more excited auroras showcase bursts of red and white lines and flares. Although rare, you may see a blue-tinged curtain of light wave and slowly unfurl in the heavens.

Auroras originate from the Sun. Intense solar storms called coronal mass ejections shoot solar material through space at over 1 million miles per hour. When some of this material slams into Earth, it excites gases in Earth's upper atmosphere, lighting them up like a neon sign. The solar material can more easily enter Earth's magnetic field around the poles, which makes auroras much more frequent occurrences in Alaska, Canada, Scandinavia, and Russia than in Middle America. Auroras also pop up around the Antarctic Circle and can be seen in Patagonia, South Africa, and Australia. These are the southern lights, or the aurora australis.

## HOW TO FIND IT

 About once every decade the northern lights can be seen from farther south in the United States. On November 5, 2001, auroras were visible across the country. From Colorado to North Carolina, from Cincinnati to Sacramento, the aurora borealis lit up the night. A few days earlier an extremely powerful coronal mass ejection rocketed off the Sun and headed for Earth. About 2 days later the solar storm broke onto Earth's magnetic field and scattered around the upper atmosphere. Particles from the Sun danced above stargazers as far south as California, Texas, and Florida. We've had weaker auroral displays visible south of the Canadian border since 2001, but for the most part you have to travel north of the 45th parallel to have even a slim chance of seeing them.

 Although astronomers can't predict what you will see locally, they can make aurora forecasts. Visit www.spaceweather.com to get aurora alerts. They will email you when the northern lights are in your area. Even if they wake you up in the middle of the night, it is totally worth it if you can experience the incredible northern lights.

# 96 AURORAS

CLASSIFICATION: NORTHERN LIGHTS | VISIBILITY: DIFFICULT

Have you ever been outside on a sunny day and seen a strange circle of light around the Sun? Or have you ever been driving home from work and noticed a little piece of rainbow hanging out in the sky to the left or right of the Sun? If you have then you have observed a sun halo and its smaller cousin, the sun dog. Both phenomena are caused by light bending through hexagonally shaped ice crystals in Earth's atmosphere and are really cool to see.

Sun dogs are also called mock suns since they can show up as dimmer, fuzzier copies of the real Sun. Sometimes sun dogs display the colors of the rainbow—from red on the inside toward the Sun, to blue on the outside away from the Sun.

When you have more ice crystals in the atmosphere, sun dogs can extend all the way around the Sun to create a full circle, or a sun halo.

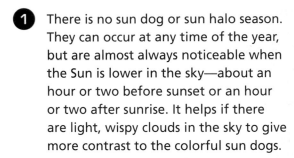

## HOW TO FIND IT

**1** There is no sun dog or sun halo season. They can occur at any time of the year, but are almost always noticeable when the Sun is lower in the sky—about an hour or two before sunset or an hour or two after sunrise. It helps if there are light, wispy clouds in the sky to give more contrast to the colorful sun dogs.

**2** Depending on the sky conditions, you may see one or two sun dogs, about 22 degrees away from the Sun either to the left or right, or both, if you're lucky. Remember never to stare directly at the Sun. However, if you see a mock sun in the sky or a tiny shard of rainbow near the Sun, extend your arm forward and spread your fingers. Cover the Sun with your thumb, and your pinkie should stretch out to touch the colorful sliver of light. That's a sun dog!

# 97 SUN HALOS AND SUN DOGS

CLASSIFICATION: **ATMOSPHERIC PHENOMENA** | VISIBILITY: **MODERATE**

When the Sun is low in the sky, light can play tricks on you. As the Sun sets lower and lower it will appear to turn yellow in color, then orange, and then red. However, at the very last second, just before it dips completely below the western horizon, the very top of the Sun can turn an eerie shade of neon green. This phenomenon is called the green flash and it is one of the most elusive sights to witness in the night sky.

At sunset, the light of the Sun must pass through more and more of Earth's atmosphere to reach your eyes. This extra atmosphere scatters the light, changes the color you see, and can even create a mirage that distorts the Sun's disc. It may shimmer as if you are looking at it through water.

## HOW TO FIND IT

**1** To see the green flash you will need impeccable timing, a clear view to the horizon, crystal clear skies, and a lot of luck. It is often easier to observe when the Sun sets above a large body of water.

**2** Keep an eye on the skyline and look at the sunset right before it slips below the horizon. The flash will be very quick and small, so be ready! If you want the best chance to see it, you can try to capture the green flash with a camera. Set your camera on a tripod, aim it at the Sun when it is almost touching the horizon, and start recording. A green flash can happen in an instant and the camera is often more sensitive to the green light than your eye. You can also play back the video later, slow it down, and increase the contrast. And maybe, just maybe, you can capture this ephemeral light show that few have ever seen.

CLASSIFICATION: **ATMOSPHERIC PHENOMENON** | VISIBILITY: **DIFFICULT (BUT COOL)**

There are few astronomical events that demonstrate so beautifully the heavenly dance of the solar system like a lunar eclipse. This rare alignment of the Sun, Moon, and Earth creates one of the best shows in naked-eye astronomy.

A lunar eclipse occurs when the Sun is on one side of us (about 93,000,000 miles away), the Moon is on the other side (about 240,000 miles away), and the Earth is right in the middle. The Moon circles directly into the darkest shadow cast by Earth into outer space, and our planet blocks the sunlight from reaching the lunar surface. From Earth it looks like a dark shadow is slowly creeping across the face of the Full Moon. When it is completely covered, the Moon turns an eerie shade of pink, orange, or red. Some people call it a Blood Moon, since the darkest lunar eclipses turn the face of the Moon a crimson color.

## HOW TO FIND IT

**1** To really take in a lunar eclipse, sit outside with friends and family and make a night of it. An average lunar eclipse lasts 3 hours, and although there are two very exciting parts to it, most of the time you can just relax while the sky show goes on around you.

**2** The best part is capturing the beginning of the eclipse—when the first bit of Earth's shadow can be seen on the Moon. Astronomers predict the start of eclipses down to the exact second, and it can be a thrill to witness the precision of their predictions and see the eclipse first commence.

**3** Over the next hour, Earth's curved shadow will slowly creep across the lunar surface. Bit by bit, the darker shadow covers over craters, seas, and other landmarks.

**4** Then comes the second most dramatic part: totality. When the Moon is completely in the shadow of Earth, it turns an astonishing coppery-orange color. The Moon does not turn completely black since light is still getting to it. Sunlight bends through Earth's atmosphere and weakly bathes the Moon in a warm, orange glow. What you are seeing is the combined light of all the sunsets and sunrises of Earth projected onto the Moon.

**5** Totality can last between 1 and 106 minutes. Be patient and watch how the light and color change during the course of totality. If you look away for a minute and then turn back to the Moon again, you will see a slightly different shade. But all good things must come to an end, and Earth's shadow will eventually pass away and return the Moon to its normal state about an hour after totality. The next total lunar eclipses in the United States will be:

- May 26, 2021
- May 15, 2022
- November 8, 2022
- March 13–14, 2025
- March 3, 2026
- December 31, 2028 (West Coast only)
- June 25, 2029
- December 20, 2029

# 99

# TOTAL LUNAR ECLIPSE

**CLASSIFICATION:** ALIGNMENT OF SUN, EARTH, AND MOON | **VISIBILITY:** MODERATE

The number one astronomical event visible to the naked eye is a total solar eclipse. This experience is so much better than anything else in our list (even seeing the northern lights) that it should have its own category. It is off-the-charts amazing!

Picture this: The Moon slides slowly in front of the Sun and at just the right moment, when you are standing at just the right place, whoosh! A shadow sweeps over you, the sky turns so dark that the stars and planets come out, and the temperature drops 15 degrees in an instant. When you look up at the sky, there is a perfect black circle where the Sun used to be encircled by a beautiful white halo. This is totality and experiencing it will blow your mind.

When viewing any solar eclipse, remember to use proper eye protection. Getting the Sun and Moon to be in perfect alignment is a rare and beautiful thing. From Earth, the Moon and Sun appear to be almost the same size in the sky. At times, the Moon is just large enough and just close enough to Earth that it can barely block out the entire Sun. The longest a total solar eclipse can last is 7.5 minutes. The precision of totality is so fleeting that most total eclipses only last a few minutes. Then the Moon slides out of alignment, rays of sunlight peek out, and totality is over. You are left in awe, wondering, "When is the next one of these?"

## HOW TO FIND IT

**1** Unlike a comet, meteor storm, or aurora, a total solar eclipse is a predictable, guaranteed wow moment. Astronomers can tell you the exact place on Earth to stand and the exact second that it will start. The only thing you have to worry about is clouds, the bane of sky watchers everywhere. The next total solar eclipses in the United States will be:

- April 8, 2024
- March 30, 2033, in Alaska
- August 23, 2044, in Montana and North Dakota
- August 12, 2045
- March 30, 2052

**2** But you don't have to wait around for just the solar eclipses visible in the United States. You can become an eclipse chaser and travel to witness other total solar eclipses visible from Patagonia in 2020, Antarctica in 2021, or Australia in 2028. It is never too early to plan ahead. After you experience your first total solar eclipse, you'll never be the same.

# 100

# TOTAL SOLAR ECLIPSE

**CLASSIFICATION:** ALIGNMENT OF SUN, MOON, AND EARTH | **VISIBILITY:** DIFFICULT

# Index

Note: Page numbers in bold indicate primary discussions of things to see.

Author photo by Mary Strubble

## ABOUT THE AUTHOR

Dean Regas is the astronomer for the Cincinnati Observatory and creator and cohost of the astronomy podcast *Looking Up*. In addition to writing for *Astronomy* magazine and *Sky & Telescope* magazine, Dean is a frequent guest astronomer on the radio show *Science Friday*. He is the author of the books *Facts from Space!* and *100 Things to See in the Southern Night Sky*.